Breakthrough!

Innovation Management in Practice

Jon Glasco

GCA Thought Leader Publications

Fourth Edition | August 2017

Glasco, Jon E.

Breakthrough! Innovation Management in Practice

ISBN-10: 1-45283-080-0
EAN-13: 978-1-45283-080-3

Fourth Edition [E4.0]: August 2017

Available from Amazon.com, CreateSpace.com, and other retail outlets

Author's Notes:

Designations used by companies to distinguish their products are often claimed as trademarks. In all instances where the author is aware of a claim, the product names appear in initial capital or all capital letters. However, readers should contact the appropriate companies for more information regarding trademarks and registration.

Efforts have been made to trace copyright holders, but if any have been overlooked, the author will be pleased to take the necessary steps at the first opportunity.

Dedicated to my family

Contents

List of Tables

List of Figures

"In a world of change, the learners shall inherit the earth, while the learned shall find themselves perfectly suited for a world that no longer exists".

Eric Hoffer

Preface

In many respects, innovation is a process of exploration, learning and discovery. Virtually all organizations–including corporations, startup firms, government departments, trade associations and non-profit entities–are capable of exploring new ideas to address complex challenges and future threats and opportunities. But not all organizations succeed in crafting a *process and climate of innovation* capable of transforming ideas into new products, pioneering technologies, and innovative solutions.

My objective	To convey a practical perspective on innovation management practices; clarify the language and craft of innovation; and provide tools to manage the process of innovation. In particular, this book explores how to plan and manage *innovation projects* that involve a breakthrough or other radical innovation or disruptive model.

A successful innovation–especially when it involves a breakthrough–captures our attention. Whether it's in the newspaper, a trade journal, a presentation, white paper or other communications, the ability to innovate is portrayed as a prerequisite to prepare for and confront a host of emerging and complex challenges.

Like an explorer, the innovator often starts with an idea or vision in mind, and a map or chart showing how to get from here to there. And like the exploration into unknown territory, each innovation project is unique and involves specialists from diverse functional backgrounds, each having a personal vision or stake in the results.

But unlike the exploration–which depends on one or two strong-willed leaders with a defined mission and plan that must be implemented by a team of specialists–the people involved in innovation projects have different, often-conflicting views on how to develop the innovation, what risks to take or avoid, and what combination of technologies, processes and knowledge to exploit.

Because of conflicting views and unknown risk factors, innovation projects are by necessity chaotic. And most managers are trained to avoid or minimize chaos. It is considered a problem that should be eliminated from a well-planned and orderly workplace. Does this contribute to a good climate for work? Perhaps. *Does it contribute to a good climate for innovation?* Usually not.

If you want to develop something new–*such as a breakthrough product or solution*–then the orderly process of management may serve as an obstacle. This is especially true if the organization has a culture of "making sure that we get it right the first time." The innovation-minded leader must discover a balance between managing the more routine (standard) aspects of a business versus the less routine (non-standard) process of innovation.

Innovation is about experimentation and exploration. It means *not* getting it right the first time. Or the second. Or perhaps the third. What counts in innovation is realizing that each stage of experimentation produces new knowledge and process strengths. And this serves as a springboard to the next breakthrough. Innovation depends on a spirit of collaboration and a process that encourages experimentation.

This book is about what it takes to *plan, design and manage the process of innovation.* The book addresses some of the innovation practices currently in use, and also presents theories and concepts for the future. I have included topics that I believe should be part of the discipline of innovation management and which are not generally considered in most works on the topic of innovation.

Many researchers and consultants who specialize in creativity and innovation focus on problem solving and brainstorming that occurs within small groups and in idea generation meetings. These consulting firms offer the specialized skills of planning and facilitating meetings. Within the boundary of their services, the process of innovation in a meeting or series of meetings adds value. New ideas are generated in meetings, developed into concepts, and next steps identified. The goal is to rely on team members to carry the ideas forward to meet a goal or solve a problem.

Other innovation practitioners focus more on the macro-level picture, such as *national systems of innovation* designed to develop knowledge and improve the economy and competitiveness of a country or region. Academic researchers provide insights on other topics related to *innovation and economic growth, measurement of innovation, and networks of innovators.*

However, innovation management as a process and discipline—which should have its own set of practices and a place on the organization chart—deserves more attention.

Breakthrough! Innovation Management in Practice is based on experience, observations and knowledge gained from my work with Fortune 500 corporations, startup firms, government organizations, economic development agencies and multinational infrastructure projects. Each of these entities and projects encountered complex situations and problems demanding high levels of innovation. A few of them succeeded in meeting the challenge of planning and implementing a new solution or breakthrough.

This book is for those involved in large complex projects, strategic challenges, and the pursuit of breakthrough solutions, technologies and products. My goal is that the techniques and tools presented in this book will be useful to innovators throughout the world in commercial and public sectors.

1 | Introduction to Innovation Management

"The best way to predict the future is to create it."

Peter Drucker

Although innovation is considered an essential part of life in modern organizations, few companies or government departments have placed the role of *Chief Innovation Officer* on their organization charts. And unlike the professional disciplines of marketing, engineering, finance, accounting and human resources, there have (until recently) been few university or corporate training programs that teach the theory and fundamentals of *innovation management.*

For many, innovation has been regarded as an event or activity that happens in the white spaces of the organization chart and is performed by those unusual and adventurous employees who combine an entrepreneurial spirit with other unique characteristics (which are probably not part of their job descriptions).

The result is that organizations often view innovation as a fuzzy and difficult-to-define phenomenon which–although critical to the success of companies and the health of economies–is not a function actively managed according to an established process, industry standards or best practices.

This is in spite of the fact that extensive research has been devoted to concepts and models such as open innovation, triple helix, design thinking, innovation networks, soft innovation, disruptive innovation, and radical versus incremental innovation.

Because innovation is sometimes described as a process for *creative destruction*, companies have assumed that if they occasionally (or accidentally) hire a few creative thinkers and manage the traditional functions, then some innovative individuals will emerge somewhere in the organization, identify opportunities for innovation, and take the risks and steps required to accomplish change.

5

An issue faced by many hopeful innovators is that even the organizations which promote innovation as critical to success inadvertently erect barriers to change:

- Established processes resistant to change
- Unrealistic ROI thresholds for approval of innovation projects
- Risk-averse management style
- Lack of management commitment
- Unwilling to consider disruptive changes
- Too much emphasis on past success
- Poor understanding of market and competitive forces
- Not-invented-here mindset

These barriers prevent ideas and concepts from crossing over organizational boundaries and hinder people from venturing into uncharted, risk-filled territory where opportunities wait to be discovered. It's not that companies construct barriers intentionally. And it's not that companies want to prevent their employees from innovating. It's just that inadvertently, barriers are formed because companies and their executives do not have a process or the specialized skills needed to create a healthy setting for innovation.

This book is about what it takes to (a) create such a setting; (b) design and manage a process of innovation; and (c) integrate the process with other activities.

What is Innovation Management?

The word *innovation* is sometimes used almost interchangeably with *creativity*. However, the two terms refer to different activities. Creativity is usually a highly individual activity that involves using the imagination to generate new ideas, accomplish leaps of insight, and make connections between thoughts that were previously distinct. Innovation is a process for transforming and developing ideas into feasible and successful concepts and ultimately implementing the concept in the form of a new product, service, system or solution.

Defining Innovation and Breakthroughs

When asked to define innovation, people will often reply that this refers to a new invention, technology, product or process. For many, innovation is seen as a mysterious event, difficult to define or manage, and something that ventures into risk-filled territory they would prefer to avoid.

An expansive view of innovation is one that cuts across society, business and government problem-solving efforts. As described by thought leaders (e.g., Chesbrough (2006), Osterwalder and Pigneur (2010), Keeley et. al. (2013), Shukla (2017)), innovation comes in a wide variety, including product and service innovation, process design, marketing innovation, networks of open innovation, improvements in customer engagement, supply chain innovation, business model generation, and social innovation.

Innovation refers to how a company, entrepreneur or government entity generates value from multiple creative activities. The objective of innovation is usually to solve a problem, deliver something new for customers, communities or nations, or address a socioeconomic crisis. Because innovation is a major driver of modern economies, the factors that enable innovation are critical not only to corporations but also to government leaders, urban planners, designers of public services, educators and elected officials.

Breakthrough innovations refer to the changes that produce wide-ranging and usually long-term consequences. In general, a breakthrough innovation is one that removes or leapfrogs barriers to change. A breakthrough usually requires more than an isolated discovery. It requires the discovery of an opportunity plus the emergence of a vision, a decision to place resources at risk, and a project or process leading to implementation of the vision.

Although breakthroughs do not happen every day, the opportunities for developing and implementing new-to-the-world solutions (in business, government, economic development and society at large) surround us.

Defining Innovation Management

The above definition places innovation in the context of end products or solutions. An innovation is what happens at the end of a long and complex process, starting from a rough idea or concept. Innovation is accomplished when an organization makes a decision to pursue an opportunity, commits resources to a process, and manages the process to produce desired results.

7

However, this book is concerned with more than the end result of innovation. It is about *how to get there*–i.e., how to design and manage the process; develop resources and tools required for innovation success; develop breakthrough solutions; and capture the value. In this book, *innovation management* is therefore defined as follows.

Innovation Management	The process, tools, climate and resources needed to generate, nurture and develop new ideas, concepts and solutions–and derive value from the results

With this definition, we look beyond individual creativity to the more complex landscape of *managing the process of innovation*–which deals with the pursuit of opportunities. Although an idea is always at the nucleus of an opportunity, the idea has little value until it is transformed into a workable concept, developed into a prototype, tested, produced and introduced in a market or situation where it will succeed.

There are numerous processes and projects that demand innovation management skills in the pursuit of opportunities, for example:

- Development of technologies, products and services
- Organizational improvements and process changes
- Business model transformation
- Implementation of joint ventures, mergers and multinational partnerships
- Improvements in system integration and logistics
- Government transformation programs
- Removal of barriers to entrepreneurship and economic development
- Development of national innovation systems, technology parks, business incubators and other infrastructure and public services

Table 1-1 distinguishes between two categories of *innovation activities*–short-term change initiatives versus large-scale innovation projects and communications. Both categories may be involved in the types of innovation projects and processes in the preceding list.

8

Table 1-1. Innovation Management (IM) Activities

Attributes	Short-term change initiatives	Large-scale innovation projects
Duration	Brief	Lengthy
Importance of non-traditional roles	Sometimes important	Essential
Relies on executive decision	Usually	Always
Depends on cross-functional support	Maybe	Yes
Relies on diverse thinking	Yes	Yes
Depends on persuasive communications	Maybe	Always
Depends on trust and risk-taking	Sometimes	Always

Why Innovation Management is Important

Many organizations recognize that downsizing and other cost management tactics are not enough to ensure long-term success. These organizations understand the role of innovation as a critical part of strategy and leadership. Leading innovators outperform competitors in terms of valuation, time-to-market, and adherence to quality and cost targets. According to data from A. T. Kearney, innovation leaders have significantly higher growth and profitability than average performers (Wagner, 2007).

"In a world where industry disruption is increasingly the norm, not an anomaly, virtually no company can ignore the imperative to innovate. Failing to do so is an invitation to lose business." (Staack and Cole, 2017)

According to Sernack (2017), Knowledge@Wharton (2016), and Wagner (2007), the importance of innovation stems from the need to:

9

- Increase business growth, profitability and value

- Respond effectively and quickly to disruptive market and competitive forces

- Capture value from R&D and intellectual property

- Take advantage of advances in digital technologies and services

- Increase competitive strengths and market share

- Identify emerging customer opportunities and enter new markets

- Harness and take advantage of entrepreneurial talent (within the organization and throughout the ecosystem)

- Adapt to changes in the external environment (e.g., regulatory, legal, financial, political)

- Communicate with stakeholders (e.g., concerning business model change) and build trust

- Design and implement sustainable solutions

- Respond to complexity, risk and wicked problems

- Solve problems and add value to quality of life

According to Staack and Cole (2017), "companies applying customer-engagement strategies that employ design thinking and user-driven requirements from ideation to product/service launch are about twice as likely as their survey peers to expect growth of 15 percent or more over the next five years."

Because of increasing pressures on organizations to change and innovate, many consulting firms, universities, government entities, and trade associations have taken an active role in the field of innovation practices. Much of the research in the field of innovation examines the factors that influence risks and decision-making in innovation projects, especially those involving development and introduction of new technologies, products and services.

For a given threat, problem or opportunity, numerous sources of innovation and signals from the external environment or from within the organization may propel creative thinkers to craft an innovative vision or solution. Figure 1-1 summarizes the types of sources and challenges an innovator encounters in a 21st century enterprise.

Figure 1-1. Innovation Sources and Challenges

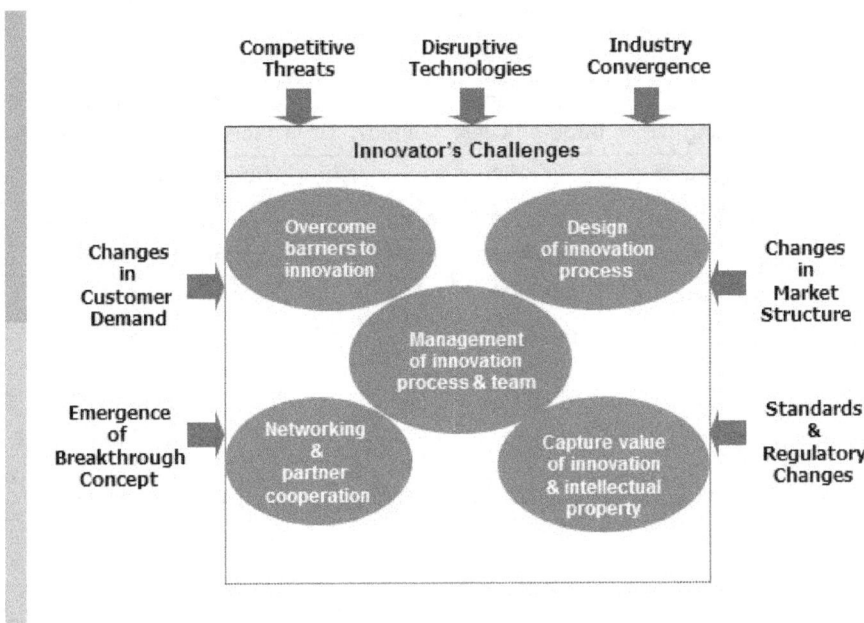

To address the challenges inherent in the innovation process, Chapter 3 provides recommendations for a process template and discusses the success factors in managing the process.

How Innovation Projects Differ from Incremental Improvement

Innovation projects differ from activities involving incremental improvements—which are often created and implemented within a single department and may not require executive approval or large resource commitments.

For an incremental improvement, the process is usually one that is proven by experience, and team members know the path toward success. However, in an innovation project, the process is not well defined in early stages, and a path to success is imprecise. Table 1-2 summarizes these differences.

Table 1-2. Incremental Improvements versus Innovation Projects

Attributes	Incremental improvements	Innovation projects
Major issues	Tactical & technical	Strategic & tactical
Technical understanding	High to very high (most trade-offs well defined)	Low to very low (trade-offs poorly defined)
Historical data	Plentiful	Sparse to non-existent
Process	Detailed & proven by past experience	Not well-defined (may change frequently)
Path to success	Understood by all	Unclear (many options)
Decision maker(s)	Identified throughout the project	Often not identifiable in early stages
Role clarity	High	Low
Exchange of new ideas	Generally low Not critical to success	High throughout project Critical success factor
Importance of persuasive communications	Low	High
Handling uncertainties & problems	Formal procedure Logical / analytical process Guessing not allowed!	Informal Not very analytical Lots of *educated guessing*
Attitude toward risk	Careful & analytical Manage the details Get it right the first time!	More intuitive Manage the process Allow for mistakes & corrective actions

Innovation projects also demand a high level of persuasion. In this respect, innovation is similar to proposal management, where the benefits and features of new products, technologies, solutions, services, and process changes are described and proposed.

An innovator needs the persuasive skills to show how a proposed innovation will yield benefits that exceed costs, how the risks of innovation will be managed, and how the innovation fits within the context of the organization's vision and strategy.

Project-Driven Organizations and Innovation Management

Project-oriented organizations have an ability to form and manage networks of individuals, many of whom come from different backgrounds and specialized fields.

Although the nature of projects varies widely, project-oriented organizations share a number of characteristics:

- Most projects are formed to accomplish specific tasks and objectives.
- Some projects will exist at the edges of the organization, working on tasks that lie outside the core business.
- Risks and rewards for project team members are not always defined or understood.
- Management practices and operating procedures are adjusted to suit the needs of the project.
- Innovation and learning is associated with the project's success, but will be difficult to transfer to the overall organization.
- Some project team members may have to assume non-traditional roles (as described in Chapter 7).
- Some team members will continue their duties in the core business unit while also devoting time and energy to the project.

Many project-driven characteristics lead to perceived risks and uncertainties for the individuals involved in the project, and thus drive the need for innovation management practices that:

- Build and maintain a good climate and entrepreneurial spirit for projects
- Provide a pool of resources trained and experienced in the process of establishing, funding, and motivating project teams
- Manage and mitigate risks to individual members
- Transfer knowledge and innovation from project teams and from external sources into the mainstream of the organization
- Identify when and where–based on project results–to make changes in the organization

13

Innovation Management: It Takes More Than R&D

> *"Research is the transfer of money into knowledge, while innovation is the transfer of knowledge into money."*
>
> UK Department of Trade and Industry

Although companies in many industries rely on R&D as a source of ideas and innovation, it is not always the first step toward becoming innovative. Reviews of available literature seldom reveals evidence that those who spend more on R&D are the top innovators in their industries.

According to PwC, twelve years of data from their annual Global Innovation 1000 study "found no statistical relationship between dollars spent on research and development (R&D) and financial performance, suggesting that the way you spend your innovation dollars is more important than how many of those dollars you spend." (Staack and Cole, 2017)

With a healthy climate for change, innovation management extends beyond the R&D function and takes place in many areas of an organization.

Innovation management is *not* synonymous with R&D management. While R&D may provide a technology foundation that enables a successful innovation, investing in R&D by itself does not always create strengths in the function of innovation management.

As discussed by Arundel, et al. (2008), "R&D is not the only method of innovating. Other methods include technology adoption, incremental changes, imitation, and combining existing knowledge in new ways." Their research indicates that in the EU, 52% of more than 4,300 innovative firms innovate without performing R&D.

In organizations that perform R&D, other functional departments are often important sources of innovation. Marketing and engineering cooperate to produce ideas and concepts for new products and services. Procurement seeks ways to improve supplier relations and quality. Strategic planning crafts a new vision and scenario—describing what the organization intends to be in the future and how it will get there.

While it is correct that the potential for innovation sometimes emerges from a company's research and development, the potential often loses momentum due to lack of communications. A new technology or concept emerges, and the natural tendency is for scientists and engineers to communicate via technical papers and conference presentations. This method typically offers

14

incremental progress in a given field, and may ultimately yield incremental improvements in a product or service. The company then tries to forecast a future need that might be met with the research results. However, groups often have a tendency to think about future changes in incremental terms. Therefore, the group's ability to create a breakthrough concept radically different from the status quo is constrained. The opportunity to leapfrog over barriers to create new-to-the-world (NTW) solutions is generally neglected. In these situations, the role of innovation management (discussed in more detail in remaining chapters) is to:

- Encourage exploratory thinking as a way to leap over barriers
- Design and facilitate a process of innovation
- Facilitate a breakthrough vision and shared paradigm of success
- Clarify the importance of non-traditional roles
- Deliver communication tools which enable customer engagement and trust
- Identify the expected value from investing in knowledge and innovation
- Develop thought leadership regarding future scenarios and *new-to-the-world* solutions

Levels of Innovation Management Projects

The available literature on innovation and related topics reveals there are many levels of innovation and development of solutions–ranging from supranational innovation programs (cross-border strategies on a multinational scale, designed to benefit multiple national and regional economies) to public / private initiatives (usually large projects designed to modernize infrastructure or develop a new technology) to the activities of startup firms trying to develop and commercialize a new product. Figure 1-2 depicts the range of innovation management levels–from large multinational projects to incremental improvements.

Figure 1-2. Examples of Innovation Management Levels

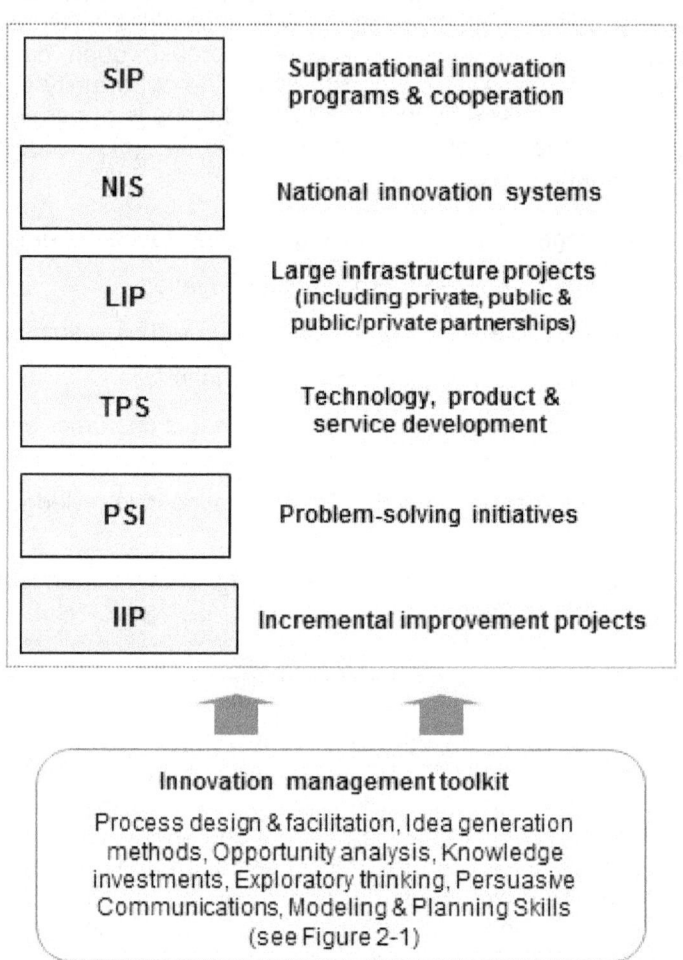

In this book, my focus is on innovation that occurs at the macro levels (MIP, NIS, LIP and NPS) projects. However, many of the concepts and principles apply to all levels of innovation, and breakthrough innovations can be implemented at any level.

2 | Innovation Management Tools

> *"People are inherently creative. They will use tools in ways the toolmakers never thought possible."*
>
> Steve Jobs

It's a long journey from an idea to a successful innovation. When you decide to form a specialized team to explore a new opportunity or address an emerging threat, your team must deal with the challenge of generating ideas; selecting ideas for further development; transforming ideas into feasible concepts; securing financing and other resources for innovation projects; and ultimately designing and implementing new solutions, products, and services. What innovation management tools and techniques are needed for this team and process?

Based on my research, participation in many innovation projects, and observation of successful innovators in several industries, the tools for innovation management include:

- Process design and facilitation
- Innovation networks
- Knowledge investments
- Discovery-driven opportunity analysis
- Persuasive communications
- Modeling and planning systems

It is probable that your organization already has people with skills in some of these areas. But the key to successful innovation is to ensure the team of innovators has all the tools available (or can acquire them) and the ability to apply the tools to the challenge.

As depicted in Figure 2-1, an innovation management (IM) project relies on having the tools in place and a good climate and culture for innovative activities.

Figure 2-1. Innovation Management (IM) Framework

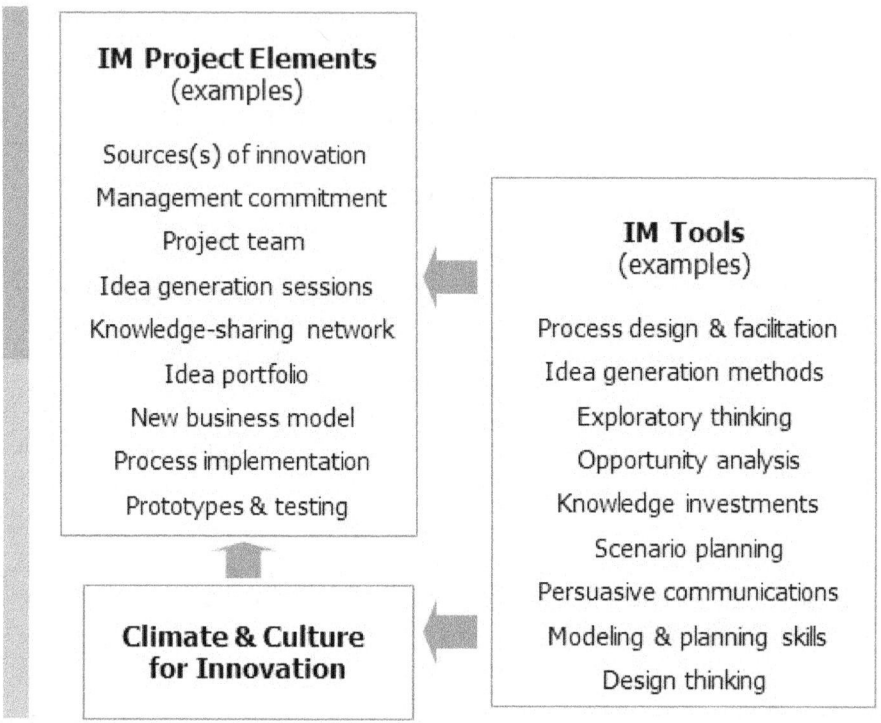

Process Design and Facilitation

Similar to other work, innovation is a process. But unlike other forms of work–where the process is usually established through results and acceptance, eventually standardized, and applied to the work in an orderly manner–the process of innovation is often created to suit the project and *may not be orderly!*

This results in a requirement for a unique form of process management. In an innovation project, the leader–by working closely with a facilitator–must ensure all members of the team understand and commit to the process and know where they are in the process at any given time. Because the process of innovation may change during the project, this makes the leader's job more difficult. Some team members become comfortable with the initial process (similar to their acceptance of other, established processes). Handling the discomfort caused by process change is part of managing and facilitating the process of innovation.

Another aspect of innovation projects is that in some cases, a vision of a breakthrough begins with a vision of the process, which might consist of moving along a technology development curve from point A to B and ultimately to point H. However, if someone on the team sees a way to get from C to H directly (leapfrogging over D through G), then the leader must revisit the process diagram and prepare the team and management for a more aggressive approach.

In summary, the process skills needed on an innovation project include:

- Ability to manage the creation of unique processes (and to define the linkage with established processes and departments)
- Diplomatic and communication skills in presenting a new process to management, team members, partners and other stakeholders
- Ability to identify resource requirements in each phase of the process and to allocate resources effectively
- Skill in solving process problems
- Flexible style that enables changes in the innovation process
- Process facilitation strengths (skill in fostering collaboration among team members, facilitating idea generation and problem-solving workshops, and helping team members to communicate)

Innovation Networks — Formal and Informal

When people work together to accomplish something unique, they must collaborate and communicate. This may happen on a one-on-one basis or in small or large groups. Communications may take place at one or many locations or between locations. The process of innovation is like an engine. To provide fuel–i.e., knowledge–for the engine and to keep it operating without breakdowns, the innovation team needs to create a network of interaction and maintain a dialogue of knowledge sharing and problem solving.

Although people who join an innovation team have their own professional networks (and these come into play during the project), they will also rely on an informal *network of innovators* characterized by interactions concerning:

- Project objectives, priorities, success factors
- Process management plans, schedules, resources

19

- Knowledge sharing and ideas on market opportunities, competitive dynamics, cost models, design prototypes
- Analysis and expert opinion on legal, regulatory, environmental, safety, standards and product usability issues
- Project team roles and responsibilities (including recognition of non-traditional roles)
- Persuasive communications (e.g. how to secure approval of funding for a new technology, business model or proposed design)
- Conflict resolution and how to overcome barriers
- Project transitions and handoffs of interim results

Knowledge Investments

An intrinsic quality of innovation projects is that existing knowledge is usually not enough. Somewhere along the path to successful innovation, existing knowledge may be challenged or deemed insufficient. Therefore, the project leader and sponsor should be prepared to invest in obtaining new knowledge in the form of research and analysis on:

- Technology developments relating to the innovation
- Market and customer trends that point to shifts in demand or requirements for unique solutions
- How the innovation will be integrated within the company's operations (or placed in a separate business unit or subsidiary)
- Risks of implementation (safety, environment, financial, quality)
- Potential partnerships (e.g., What partners are needed in the innovation process? What is their role and value?)

By necessity, the innovation manager must be aware of the nature of the company's internal knowledge networks. Because ideas, creative thinking, new product concepts, process improvements and other proposals emerge from numerous internal sources, the role of innovation management should be networked with the role of intellectual property manager.

The principle of open innovation (Chesbrough, 2006) reveals that the perspective you need to identify an emerging opportunity comes not only from internal sources but also from partners, suppliers, customers, and other external sources. The value of an innovator's knowledge network is often based on the diversity of knowledge transfer sources–such as professional societies, social media networks, colleagues with unique expertise, consulting and market research firms, supplier contacts, investors, writers, trade associations, university contacts, industry events, and collaboration with peers.

In many situations, the innovation manager is concerned with the process and time required to accomplish knowledge turns. As defined by Dr. Andrew Grove (1998), a knowledge turn is the time required for an idea or experiment to proceed from initial hypothesis to results in the marketplace (about 18 months in the semiconductor industry, but more than 10 years in other sectors, such as pharmaceuticals, public transport systems, and aerospace).

Discovery-Driven Opportunity Analysis

Although innovation projects are sometimes launched to solve a problem or reduce a threat, most visionaries, entrepreneurs and innovation-minded leaders are motivated by opportunities and the potential value of pursuing the opportunity. As Peter Drucker discussed in *Innovation and Entrepreneurship* (Drucker, 1985), opportunities for innovation emerge from many sources.

Whether you apply traditional brainstorming, white space mapping or other idea generation methods, opportunities often emerge when someone discovers an idea, unique design or potential solution at the intersection between two or more diverse fields of activity. An example of this comes from the intersection of public healthcare services and micro-finance services— which led to new ideas to address the global challenge of providing healthcare services in underdeveloped countries.

Another example comes from the success of Michael Dell, who envisioned the combination of computers and mail order services and formed Dell Computer Corporation. This is what Arthur Koestler (1990) refers to as seizing an opportunity through *bisociation*—the "ability to relate two seemingly unrelated things to produce the *ah-ha* sensation in the market." (Smilor, 1996)

Unfortunately, the tendency for some decision makers is to reach judgmental decisions or evaluate emerging opportunities with an analytical mindset that worked in the past (but does not apply to the future). This results in harsh treatment of new opportunities–by creating unrealistic barriers to innovation. The mindset to evaluate a proposed innovation must sometimes shift from an old analytical model to a new model that provides a balanced assessment and encourages fresh thinking with intuitive and analytical points of view.

A balanced opportunity assessment recognizes the importance of potential rewards and risks for the innovator. Most people do not enter the complex process of innovation simply because it is part of their jobs. For many innovators, the process of discovering new opportunities is related to the tradeoff between risk and reward. This suggests the importance of offering an equitable system of rewards and demonstrates the power of motivation in the discovery process.

However, the would-be innovator or intrapreneur may focus *only* on discoveries that offer potential rewards. If the defined rewards are the dominant motivator and the organization has a risk-averse culture, then the possibility of discovering new-to-the-world solutions (outside the mainstream of the organization or project) is diminished.

Therefore, the organization needs an approach to risk-taking and rewards that allows for assessment of a wide range of opportunities, including those within the mainstream business and those outside. Opportunities outside the mainstream require different analytical models and new ways of thinking.

Regardless of the methods or motivation for discovery of opportunities, two notable characteristics of the opportunity-driven innovator are *exploratory thinking skills* and the *ability to shift from exploratory to analytical thinking* at the appropriate time in the process.

Exploratory Thinking Skills

Most technical and business people learn at an early stage in their training what it means to apply the scientific method to challenges. In general, scientific thinking skills involve achieving clarity in the mind as to purpose, problems and methods. Scientific thinkers tend to question much of the information and conclusions presented to them. They are logical and objective in how they think about issues and challenges.

However, in the context of innovation, I propose a definition of *exploratory thinking*, one that involves less reliance on logical evaluation and less questioning in the early phases of the innovation process.

Exploratory Thinking	An approach to thinking intuitively about challenges that demand change, problem-solving and unique solutions -- while encouraging innovators to *avoid go/no go evaluation* of new ideas; building upon the ideas of others; and considering the positive aspects of emerging ideas

Nobel Laureate Herbert Simon believed that innovative people develop "chunks" of knowledge (Simon, 1985). These are "sets of patterns and relationships that develop over time and allow one to see solutions to problems—by making connections between events and actions" (Smilor, 1996). This is a form of exploratory thinking as defined above. An entrepreneur develops *chunks of opportunity-centric information* through professional networks, exploratory market research, interaction with customers, knowledge of market and technology trends, and an understanding of competitive dynamics and regulatory issues.

Shifting from Exploratory to Analytical Thinking

Conducting a thorough analysis of the opportunity, while adhering to a new model and its metrics, is where members of the innovation team must depart from the exploratory thinking methods they used earlier in the process.

At this point, a decision maker must look objectively at the proposed innovation, and decide whether to continue and how to proceed with the project. In most situations in a corporate setting, this is where analytical skills and tools must be applied. However, it is important to recognize that because most innovations involve some degree of risk and uncertainty, the true nature of the opportunity and its potential benefits may be obscured by *group think* and other organization and cultural dynamics.

Therefore, the innovation manager and other members of the team may need to intervene at various points in the analytical process to modify the concept in some manner and improve its chances of acceptance. Their challenge is to help the organization avoid Type II errors, i.e. rejecting innovations that have high attractiveness and high likelihood of success (although the attractiveness is difficult to recognize in early stages).

As taught by Chesbrough (2006), some firms have a tendency to focus on avoidance of Type I errors, i.e., approving innovation projects that have little chance of success (and which usually have high failure rates). However, the same firms will have less interest (or ability) to avoid Type II errors. A classic example of a Type II error was when Xerox—which developed technologies for the personal computer and mouse—decided not to proceed with commercial development of these products (due to a corporate view that these were not a good strategic fit with their core business). IBM's decision not to secure an exclusive license to Microsoft's operating system was a Type II error with long-term impact. Perhaps another example of a massive Type II error was in the late 1940s and 1950s when US leaders had an opportunity to create a program of national health services for low-income families and failed to take action.

Funding issues are often the most difficult challenges to overcome in innovation projects. Therefore, the innovation team may need to conduct workshops in parallel with the opportunity analysis in order to control the expected cost burden–and craft a new financing strategy.

Persuasive Communications

> *"Your effectiveness depends on your ability to reach others through the spoken and written word."*

> Peter Drucker

Most innovations begin with an idea or group of ideas that connect in some manner to yield a vision of a breakthrough or new-to-the-world solution. Ideas are then refined and developed into an attractive concept, eventually leading to a plan or proposal to secure management commitment and resources.

The vision of change–and the potential benefits of an innovation–are often crafted in imprecise terms and are unproven in the early stages. It is the quality of communication and its persuasive impact, usually in combination with the credibility of the project leader and sponsor that secures approval for the project.

Many large projects depend on the skills of a technical communicator to document the project plans, proposals, and findings. Both technical communications and persuasive communications are necessary when developing something that is new and unique. However, the difference is that persuasive communication methods appeal to intuition and beliefs and are *influence-driven* whereas technical communication methods are *information-intensive and data-driven*.

Having persuasive communication skills in the innovator's toolkit—to influence decision-makers, investors, team members and partners—is a key element of success on innovation projects. Chapter 6 discusses the principles of persuasive communications.

Modeling and Planning Systems

As in most other projects, the innovation project must at some point deliver a credible analysis and forecast of value, benefits, costs and risks.

Breakthrough! Innovation Management in Practice

Although the facilitator should encourage innovation team members and management to avoid judgmental thinking in the early phases of innovation, the fact of business life is that somewhere in the innovation process (before approval of resources), the team must present a favorable financial forecast.

However, the innovation project leader and sponsor need to ensure the model and analytical methods applied to the project are appropriate to the situation. Analytical techniques used in the existing business—to evaluate incremental improvements—may not be relevant to a proposed innovation that involves a breakthrough or radical change from the status quo. In this situation, the innovation sponsor will confront several questions:

- What model and analytical approach are needed?
- Who can develop a new model to evaluate a breakthrough?
- What new variables and success factors should the model address?
- How should we present the model to management—as a means of building credibility and commitment?

Shared Paradigm of Success

Another aspect of the analytical challenge is that executives involved in the core business will each have a mental model of the business and its strengths, weaknesses and opportunities. In most cases, these intuitive models have subtle differences from the accepted spreadsheets. The *accepted spreadsheet model* may not address the intangible features of the market and competitive forces, which the mental models attempt to do. The facilitator's role is to help the team craft a *shared paradigm of success*, based on the vision, mental models, and an appropriate analytical model. This involves the following activities:

- Articulating the vision from a project sponsor or champion
- Ensuring the opportunity (or threat) and vision are communicated to team members and stakeholders
- Giving team members a climate and informal network for exploratory thinking and idea generation
- Facilitating the creation of an intuitive model for the project
- Retiring old analytical models not relevant to the project and ensuring a new analytical model is developed
- Communicating with stakeholders to clarify the expected outcome (a new paradigm)

25

Figure 2-2 depicts the dynamics of crafting a shared paradigm of successful innovation.

Figure 2-2. Shared Paradigm of Successful Innovation

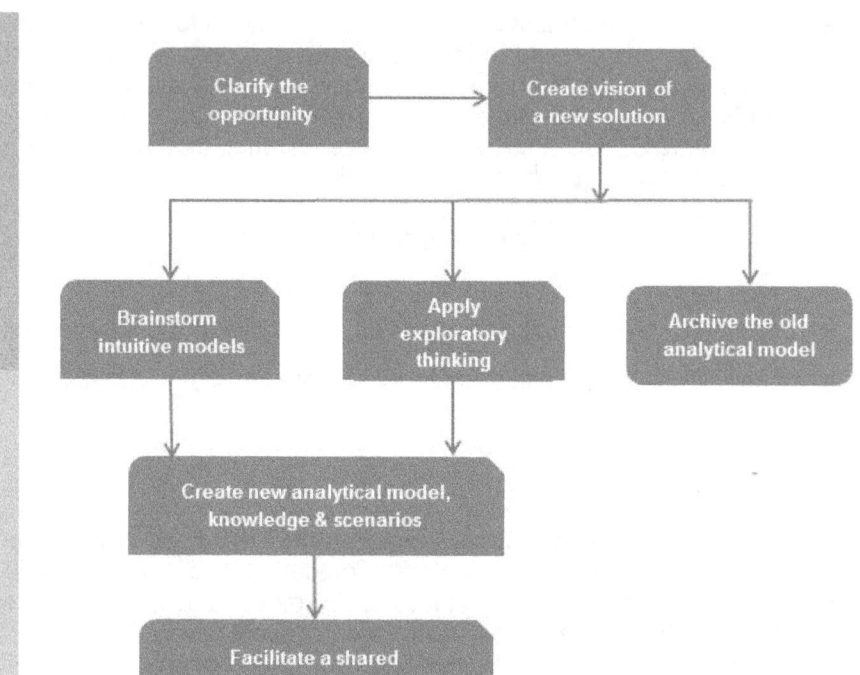

If decision makers perceive that a proposed innovation project is close to the mainstream business, they will apply their existing mental model and the accepted spreadsheet to the proposed innovation. This is usually bad news for the innovation project leader, who needs to be persuasive in presenting and recommending a new model and analytical framework relevant to the innovation.

Paying the Penalty of an Unsuitable Model

When I was involved in strategic planning in the US telecommunications industry, I was fortunate to work for a senior executive who was highly receptive to ideas and concepts for new products and services. He was also very knowledgeable of all aspects of the telecommunications business, from technology planning to operations to financial management, and could

Breakthrough! Innovation Management in Practice

evaluate a proposed product or service by relying on his intuitive model and then mentally estimating expected revenues, costs and profitability.

When a new business concept (outside the mainstream telephony business) was presented—that of mobile cellular phone services—the mental model and existing spreadsheets no longer applied. When the executive sought advice on the potential for this business, he was told that "cellular services would only be used by business customers" and there was "no attractive consumer market" for these services. This lack of vision among the company's executive team and their unwillingness to consider the emerging mass-market opportunity was unfortunate and failed to recognize the bodies of research, knowledge and marketplace signals that existed at that time.

The proposed innovation—to develop and deploy mobile networks, devices and services–went no further in that company. This was an example of the innovation team not having an appropriate model *or* sufficient persuasive skills. Needless to say, other companies recognized and pursued the massive global market opportunity for mobile telecommunications.

Breakthrough! Innovation Management in Practice

3 | The Process of Innovation Management

> *"Everyone has an idea. But it's really about executing the idea and attracting other people to help you work on the idea."*
>
> Jack Dorsey

Most people who have been involved in development of a new technology, product or system will confirm that innovation is an exciting pursuit (especially if it's a successful project). However, they will also tell you it's often hard work. And all work is a process.

Some organizations committed to innovation eventually develop a process unique to their industry, customers, internal culture and strategic objectives. Investors, business partners and other external stakeholders also exert pressures to innovate.

While it's difficult to explain how the sparks of genius in a lone inventor produce new ideas, I have observed in many organizations that the collaborative process of innovation is definable in a general way. Working on a tough issue (the *issue identification* phase) leads to a wide range of ideas generated by a group (in the *ideation phase*). The initial seeds of fragile ideas are developed into a more concrete concept (in the *concept / prototype development* phase) which is analyzed and evaluated for feasibility.

In the next phase (*project selection*) a decision is made to fund the innovation project. The prototype is transformed into a *design version* that undergoes user testing. Based on test results, the design is modified, documented, and ultimately introduced to a target group of users (the *deployment* phase). This seven-step generic model is helpful as an innovation process template:

Innovation Management (IM) Process Model

I	Planning and formation of innovation team
II	Issue identification
III	Ideation
IV	Concept / prototype development
V	Project selection
VI	Design & test
VII	Deployment

Figure 3-1. Innovation Process Template

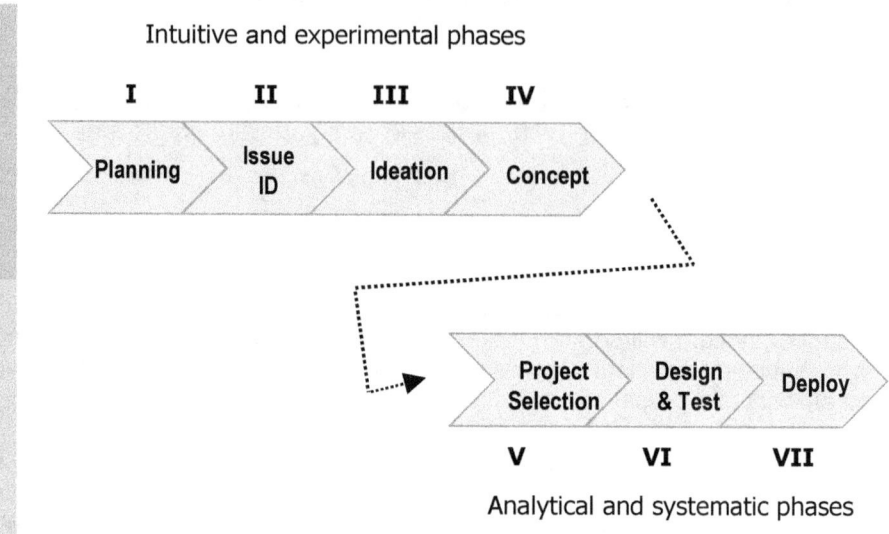

Intuitive and experimental phases

I   II   III   IV

Planning — Issue ID — Ideation — Concept

Project Selection — Design & Test — Deploy

V   VI   VII

Analytical and systematic phases

The first four phases are the *intuitive and experimental* phases. These phases often cause difficulty and tension for project teams, because there is a tendency to begin evaluating new ideas too early. The intuitive and experimental phases are designed to avoid (or at least postpone) evaluation and judgment. There is less emphasis on analysis in these phases, although this is a challenge for those who expect to see the quantitative analysis of costs, financials, return on investment, and other traditional metrics.

Phase I–Planning

Generally a task force (the *Innovation Team*) with management endorsement and high visibility is formed to design and manage the process of innovation. Members of the team must have an understanding of their roles and responsibilities, which differ from their traditional duties and job descriptions.

In general, every innovation project or team relies on an owner or sponsor. This is someone who provides the initial impetus for the project and perhaps a vision. It is usually someone who holds a leadership position in the organization and has access to the resources needed to get the project approved, funded and staffed.

In addition to the technical staff and other content experts, the innovation project needs someone to serve in the role of facilitator. For virtually all

30

innovation endeavors, this role is indispensable. The facilitator is a process specialist who is not involved in the content of the project, but instead acts a guide to the process of innovation.

In this planning phase, one of the first steps is an articulation of the process in a manner that ensures all team members understand the process and their role in each phase. In most situations, a detailed project plan is not needed at this point.

Other planning actions should include:

- Selection of the initial team members (many of whom will be involved only in the early phases of the process)
- Articulation of a draft vision and mission statement
- Development of a project schedule
- Clarification of what is expected in Phase II, Issue Identification
- Coaching the team members on their roles
- Creating a climate favorable to innovation, quality of communications, knowledge sharing, and trust

Phase II–Issue Identification

Clarifying the issues that drive innovation is fundamental. An issue can be a new opportunity; a threat to the organization or its customers, suppliers or partners; or a tough problem (regarding costs, quality, staffing, safety, or operations continuity) that has not been solved through conventional practices. The information or knowledge leading to issue identification may come from either external or internal sources of innovation.

It is not always necessary to define the issues in highly precise terms–or to spend large amounts of time in problem definition activities. It is usually sufficient to identify and describe issues in terms of actionable *"How to"* *statements*, for example: "How to improve safety for rail passengers at the same time that we add high speed services, new technologies, and more capacity."

Issues and problems are usually multi-dimensional, and allowing team members to have their own view of issue identification often provides unique insights and ideas.

31

Phase III–Ideation

Ideation, also known as the *idea generation* phase, is designed to obtain a wide range and large volume of fresh ideas that address the opportunities, challenges and issues.

A variety of proven approaches (brainstorming, Synectics®, scenario planning, storyboarding) are available to generate and nurture the ideas. This is a creative activity (rather than analytical) and is highly dependent on having an internal climate conducive to expressing new ideas and concepts in a collaborative manner.

There should be minimal attention to analysis of ideas in this phase. Feasibility of implementation may appear difficult and risky for many ideas. However, deciding on feasibility at this stage should not be an objective. Instead, the goal is to generate new ideas, even those in embryonic form. Evaluating feasibility comes later. The innovation team is not in a decision mode in this phase. Instead, they are in an exploratory or developmental mode of thinking where they should defer judgment of the emerging, fragile ideas.

The organizational effort required to suspend judgment until later makes this a difficult phase for many organizations. This is partly because people have a tendency to look first at the potential shortcomings and risks. Acting on their own initiative, they assume the role of devil's advocate when this role is not needed. What is needed is a mechanism and climate for protecting and nurturing the ideas.

Protecting the idea generators and the ideas through deferral of judgment is a key role of the facilitator. This is especially true on large complex projects that involve collaborative partnerships such as building design and construction, new infrastructure for public transport systems, defense projects, and design and planning of telecommunication networks. It is a characteristic–almost intrinsic to such partnerships–that in the early stages the partners may not have high levels of trust.

This manifests itself in a reluctance to share information, knowledge and ideas for solutions. Engineers on such projects are frequently in a problem-solving mode due to the unique requirements and idiosyncratic nature of each customer. In some countries, it is part of the professional protocol for technical personnel to not state their views during meetings when partners are present. This is a *cultural tendency* that relates to the need to avoid losing face.

But in the early stages of a design activity, the team needs to discover and discuss ideas. The facilitator's role—with commitment from management—is to create the conditions that stimulate and motivate team members to express their ideas and emerging thoughts about the challenges.

One way to accomplish this is to create a process that allows the team members to submit ideas anonymously. The facilitator should encourage the team to discuss the positive aspects of selected ideas (with questions and perceived negative features held in the background) before moving on to another task and saving this group of ideas for later evaluation. This gives team members time to revisit the ideas, submit further thinking, and build upon the ideas they consider attractive.

Ultimately, some quantity of ideas must be selected for further development. The selection criteria for development vary from one situation to another, but generally revolve around expected value to the organization, its goals, and its customers.

Discover and Nurture Fresh Ideas

Various methods of idea generation—such as Synectics® and other techniques—may be used to generate and nurture fresh ideas. This is a highly creative step dependent on creating an internal climate that encourages new ideas and concepts. There should be little attention to the feasibility of ideas at this point. Instead, the goal is to get new ideas on the table. The fact that *roughly defined* ideas are encouraged presents a problem similar to language barriers—because the words needed to characterize an emerging idea may not be readily at hand. Management has to provide the facilitative resources to help contributors discover and nurture ideas.

A perilous challenge is deciding which ideas to develop in the next phase. By their nature, emerging ideas are not well-defined and not easily analyzed using conventional methods and models.

Therefore, management has few facts or information in this phase on which to base decisions. Mental and verbal models—and the readiness to make

33

decisions in spite of imperfect information—are more important than the mathematical model at this point.

Phase IV–Concept Development

Some innovation practitioners believe this part of the process is for those with an inventive and entrepreneurial spirit. But the fact is that most employees who enjoy making changes and seeing the positive results of change can learn the tools needed in this phase.

The objective of this phase is to begin transforming new ideas into realizable (and usually fundable) concepts and find ways to experiment with the concepts—perhaps through the use of models and prototypes—and build a shared paradigm of success.

The Challenge of a Shared Paradigm

Building a shared paradigm is a key role for the facilitator and owner at this stage. Use a model as a tool in creating a shared paradigm of innovation success!

The experiments and modeling activities should enable the innovation team to evaluate feasibility while controlling costs and risks. The goal is make the ideas more concrete and capable of being analyzed so that a decision to implement can be made without taking major risks.

If a concept (or group of related concepts) looks attractive, then a formal opportunity analysis may yield information about how to proceed. This exploratory process should enable management to select ideas for the next stage of feasibility testing, business case, prototyping and testing.

As shown in Figure 3-2, success in the intuitive phases (I through IV) depends on quality of communications, trust, persuasive skills and role clarity.

Figure 3-2. Success Factors in the Intuitive Phases

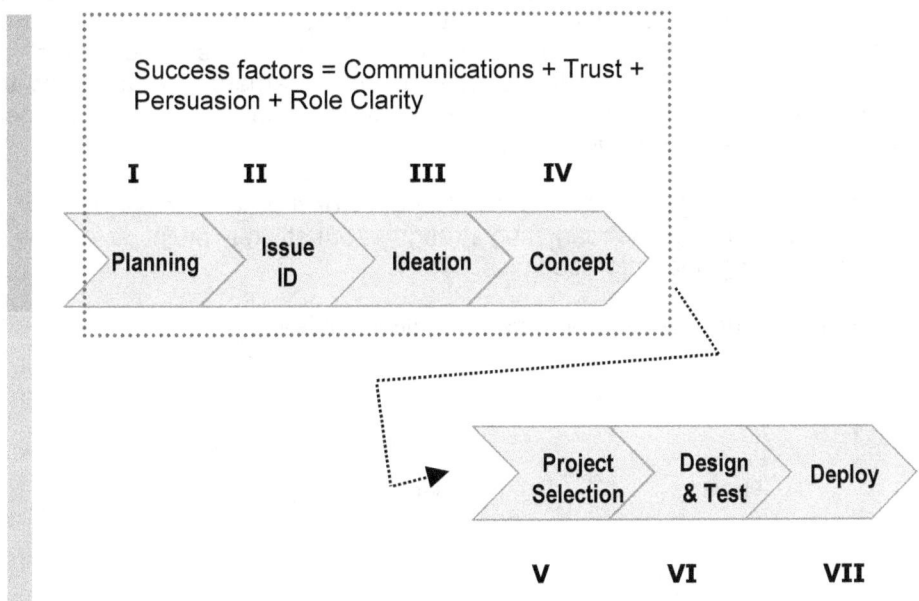

Success factors = Communications + Trust + Persuasion + Role Clarity

I   II   III   IV

Planning   Issue ID   Ideation   Concept

Project Selection   Design & Test   Deploy

V   VI   VII

Phase V–Project Selection

This is where decisions are made about whether and how to proceed with the innovation. Typically, the *go / no go decision* is based on numerous considerations (unique to each organization and situation) and analysis of strategic fit, expected value, costs, financials and deployment risks.

Prototypes and models can reduce the costs (and risks) inherent in this phase. When possible, the innovation team should take advantage of opportunities to construct multiple prototypes.

If management reaches a decision to go forward with the innovation, then traditional project management skills are applicable and should be integrated with this phase and the remaining project phases.

In some cases, members of the innovation team become members of the project management group and continue with the innovation toward test and deployment.

Phase VI–Design and Test

In this phase, the innovation begins its transformation toward a functioning product or system. Modeling skills continue to play a role in this phase if there are opportunities to test a model rather than a full-scale mockup of the product or system. If technology hurdles or other barriers emerge during this phase, then it may be necessary to return to an earlier phase to revisit ideas and concepts and make revisions.

As depicted in Figure 3-3, Phases V and VI of the IM process depend on analytical, modeling and marketing strengths (but still rely on intuitive skills for problem solving and unforeseen issues).

Figure 3-3. Success Factors in the Analytical Phases

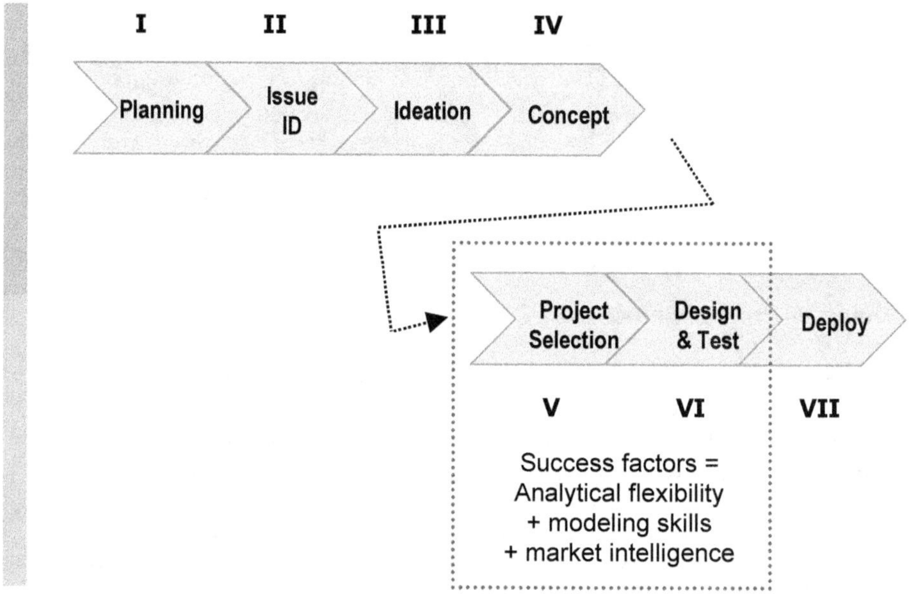

Phase VII–Implementation and Deployment

In the deployment phase, the team reaches a point where one or more concepts (after evaluation, modification, and testing) are about to become reality. This step is accomplished in various ways, depending on the nature of the challenge, the internal culture, time and budget pressures, and other project priorities.

Experience shows that implementation of a new concept or technology is usually not a smooth process. Problems will inevitably surface that demand skills in the art of creative problem solving. Therefore, some of the same techniques used in idea generation are also required in this phase.

Figure 3-4 shows that success in deployment depends on problem-solving skills, project management discipline, and marketing and communications strengths

Figure 3-4. Success Factors in the Deployment Phase

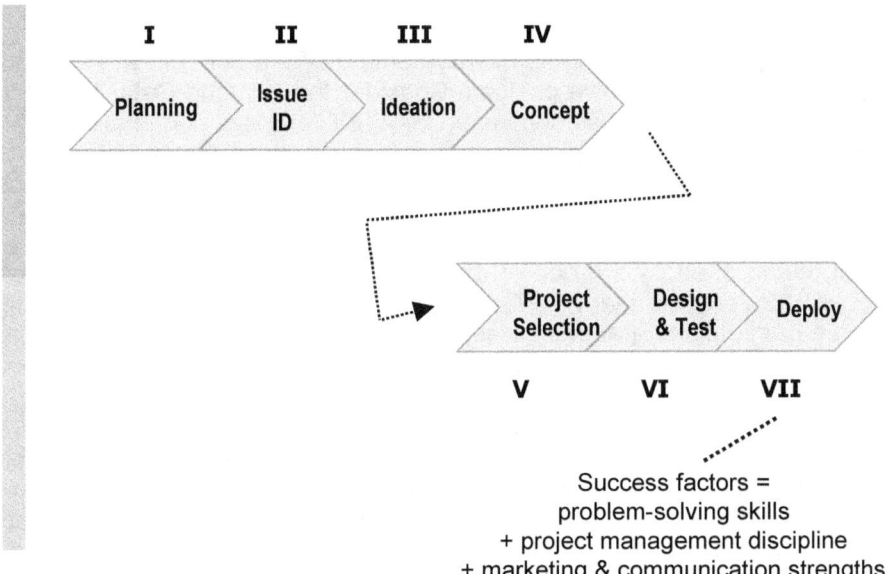

An Example of Successful Innovation in the Telecommunications Sector

Because it deals with more definable objectives, deployment is a phase of innovation where many telecommunication companies display expertise. Senior marketing managers or corporate program managers often define and coordinate the activities in this phase, with contributions from network planning, IT, customer service and other departments.

Although there are few *innovation management departments* in telecommunication companies, and the concept of innovation management is relatively new (apart from R&D departments), cross-functional and intra-industry strides are being made.

Breakthrough! Innovation Management in Practice

During an era that spanned decades, telecommunications operators developed the skills, processes and partnerships that enable success in *gradual infrastructure innovation*–such as designing and implementing a network, operations support system (OSS) and related technologies.

However, infrastructure innovation is not necessarily the best process for creation of new services, especially those oriented to differentiation and customer-centric solutions. The world's telecommunication incumbents and other operators, after developing capabilities in large, engineering-intensive infrastructure projects, were not prepared to move into an environment involving rapid deployment of new services designed to yield a competitive edge.

Opportunities for innovation sometimes emerge from shifts in market power and changes in competitive position. Consider the US long distance telecom services market, which was dominated by AT&T for many decades in the 20th century. By the 1980s, aspiring entrepreneurs realized that AT&T's position was vulnerable, and new competitors such as Sprint and MCI (who both offered discounted long distance services) seized this window of opportunity.

Sprint (established by a group of executives from Southern Pacific Railroad) and MCI understood that AT&T's long distance prices were artificially high, and the marketplace gradually accepted the risk and lower quality of the new entrants in order to achieve savings of 25 percent or more on long distance bills.

Of course, the concept of using price discounts as a market entry and share-building mechanism was not new. The issue that served as a driver of innovation was that the new competitors did not have access to the latest telecommunications technologies and equipment—which were controlled mostly by the Bell System through its dominant, monopoly position in the market. Therefore the new entrants had to build national networks using whatever telecommunications technology and equipment was available to them.

When I was a member of the strategic planning staff at Sprint during its startup years, we recognized the importance of building market share in this fast-changing market, although quality-of-service problems were challenging at this stage in the game. To begin building a national network, we relied on the microwave routes and capacity and other resources of our parent company Southern Pacific.

Finding network switching systems and other telecommunications technology to deliver long distance services was a problem, and we had to rely on off-the-shelf equipment modified to suit our needs. Designing and implementing a

Breakthrough! Innovation Management in Practice

billing system was another big hurdle in those days, and the fragile system we implemented was prone to error and loss of billing data.

Despite these barriers, we continued on the path of building market share until we established a sustainable position as the third largest long distance provider in the US market. These actions—and those of MCI and other operators—led to momentous changes in the US long distance market and a steady decline in AT&T's dominant market share position.

A Poor Climate for Innovation = Missed Opportunities

In less than a decade, the US telecom services market would face another disruptive threat—resulting from growing popularity of online services, increased use of non-voice communications, and the emerging mass-market demand for broadband capabilities. The changing market also offered abundant opportunities to develop and introduce new services.

However, many telecommunications companies had difficulties in recognizing the new reality, mainly due to internal barriers to innovation. In the early 1990s, executives in a major US telephone company (telco) were convinced that (a) "our customers use a telephone network for one purpose—to make voice calls," and (b) "this will not change in the near or distant future."

This was an unfortunate and incorrect view of market and customer trends and conflicted with the growing body of research and knowledge on future opportunities. In addition to market research, numerous signals from competitors, technology suppliers, employees and customers should have revealed that competiive threats and obvious opportunities (in terms of demand for broadband, mobile and Internet services) were on the horizon.

Events in the telecommunications industry and market soon overturned the telco's narrow perspective of customer needs and preferences. This was an example of a dominant competitor's unwillingness to recognize market signals—or take advantage of financial power and competitive strengths. In striving to protect the status quo, management failed to develop a process of innovation, encourage exploratory thinking, or create a climate of innovative energy. The result was a lack of capability and incentive to shift from a supply-centric to a customer-centric mindset.

Although telephone companies (telcos) had more than a decade to transform themselves in the digital era, they were unable to cope with changes in the nature of competition and the surprising growth of innovative digital competitors offering services in the traditional telco space (e.g., fixed voice, mobile voice, messaging). "As technological breakthroughs accelerate, more and more new digital natives are entering the core telco market with innovative business models and technologies, leaving many incumbents to wonder if they can keep up or if they will be displaced." (Meffert and Mohr, 2017)

Industry analysts (Bieler (2015), Dickgreber et. al. (2015), Holt (2017)) predict that in the 21st century—unless they accomplish fundamental changes in business strategies and operating models—telephone operators are doomed to an unsustainable role of passive [or obstructive] observers. This is the impact of disruptive innovation!

Chapter 10 addresses how, in the midst of disruptive forces, to accomplish a transformation in competitive thinking and planning capabilities by integrating strategic management with innovation management.

Building an Idea Portfolio

A *portfolio* of early-stage ideas and concepts contributes to the creation of a healthy climate for innovation.

Figure 3-5. Idea Portfolio

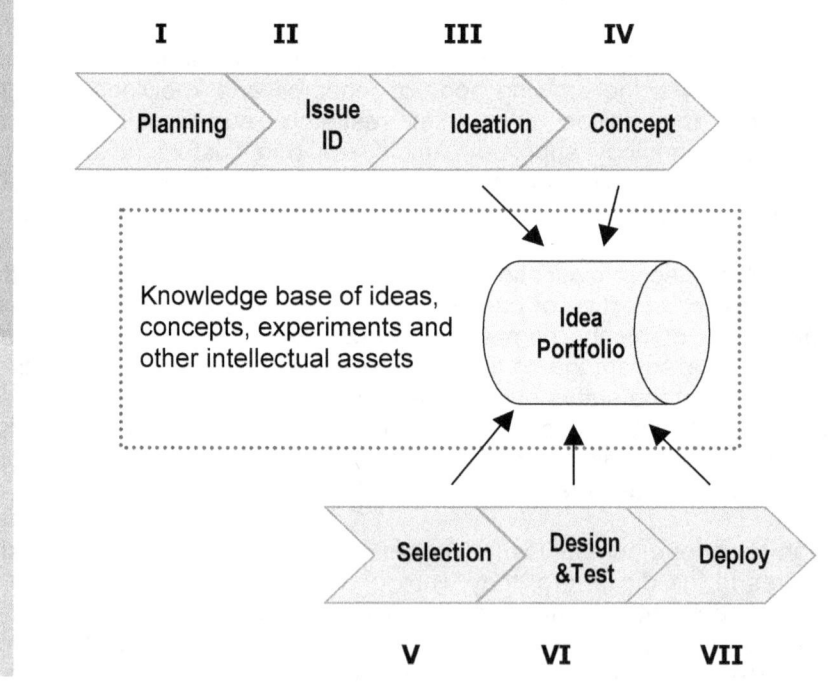

Breakthrough! Innovation Management in Practice

Through development of an idea portfolio—and encouragement of *exploratory thinking* skills*—management has an opportunity to discover a wide range and depth of ideas from varied sources, including employees, customers, vendors, alliance partners, consultants, and competitors. This adds value to the knowledge base and to the process of innovation.

Although the concept of an idea portfolio exists in many organizations (sometimes in the form of an idea pipeline), it is not something generally designed or managed in a structured manner or included in an overall innovation strategy. However, much of the potential value from innovation management practices emanates from the opportunity to capture ideas from the base of tacit knowledge and merge these ideas with explicit knowledge to create novel concepts, products and solutions.

---

* Not all innovation practitioners believe in the value of brainstorming and exploratory thinking to generate ideas and protect fragile ideas. As an example, Mark Payne, CEO of Fahrenheit 212, states that idea generation methods where every idea "is celebrated tend to fill rooms with Post-It notes rather than real value." According to Payne, "most innovations fail because their creators didn't ask tough questions at the outset." He believes "exposing fledgling ideas to the tough love of tough questions ensures those ideas can survive in the real world of real companies placing real bets with real money." (Fisher, 2014, citing Payne, 2014).

Breakthrough! Innovation Management in Practice

4 | Innovative Leadership

Based on my research and observations, every individual, group and organization has creative and innovative capacity. The innovation management challenge is to find ways to use this capacity more productively and effectively—in a manner that builds an internal climate for creativity and innovation, adds value, and contributes to strategic and operational goals.

Does Innovation Simply Happen?

It is difficult to make generalizations about how innovators succeed in accomplishing change. The process of innovation that works for a high-tech company may be quite different from that used in a publishing firm, a non-profit enterprise or a public sector organization.

However, innovation can and should be managed, like any other vital function displayed on organization charts. A quick perusal of many organization charts confirms that nearly all major companies have vice presidents and other senior executives for the traditional functions of accounting and finance, marketing and sales, operations, legal, human resources, production, and other departments. But few charts have a distinct place for the role of *Chief Innovation Officer* or a similar title.

More than once I have been surprised to hear company executives ask me "Can innovation really be managed?" The implication of course is that innovation just happens in the white spaces of the organization chart. Executives are aware that innovation is a critical success factor and that R&D, marketing, product development, and engineering are all involved in innovative activities, but the pioneering work and cooperation that must be accomplished is considered merely part of the job descriptions of leaders in those departments.

In my work—when I encounter someone assuming the role of an *innovation manager*—I observe that they often fit one of the following categories.

Breakthrough! Innovation Management in Practice

Types of innovation managers

General Practitioner–someone who understands the cross-functional nature of innovation and typically works as a program manager or facilitator to ensure the innovation process keeps moving

Knowledge Specialist–an individual with highly specialized knowledge who has the ability to see and visualize opportunities beyond the status quo

Pathfinder–offers a risk-taking mindset, sees new opportunities and is eager to create disruptive paradigms to capture the opportunity

Characteristics of an Innovative Manager

The attitude, behavior and values of innovative managers exhibit numerous similarities:

- Positive, visionary outlook
- Impatience with *business as usual*
- Commitment to an innovative culture
- Tendency to set ambitious targets
- Understanding of exploratory versus judgmental thinking
- Ability to build relationships of trust
- Persuasive communication skills
- Inclination to apply innovation management practices

A Pathfinder in the Mobile Operator Sector

A mobile operator in Austria (one of the world's most competitive mobile telecommunication markets), provides an example of the characteristics and benefits of innovative management. The company was awarded a mobile license in 1997, and arrived on the scene several years after its two main competitors had established positions in the market. It was a newcomer in this fast-growing and rapidly changing market.

The executive team recognized that time was not on their side. They knew the company's success would depend on the rapid deployment of a national network infrastructure and development of all departments and capabilities of a complete functional organization. It was time to *set ambitious targets!*

The challenge of rapid network deployment was a difficult problem. The license required them to deploy more than 900 operational sites within four months. No other operator had ever achieved this goal. Inevitably, devil's advocates advised this was not a realistic goal. However, the management team recognized this was a time for *exploratory thinking* based on a future paradigm—not judgmental thinking based on past experience. The result: they met their goal, and became the world's leader in terms of rapid network rollout success.

Using *persuasive communication* skills, this management team communicated to employees the urgent need for swift introduction of services and development of all functional departments and internal capabilities. At the same time, the company communicated to the market and customers that they were prepared to:

- Deliver customer-centric services
- Build a best-in-class mobile network
- Develop an attractive mix of services and products
- Offer a new approach to pricing and promotions
- Build a powerful national brand and distribution network
- Provide a higher level of customer service

Through their actions, the executive team demonstrated the ability to *challenge the status quo* (a dynamic market dominated by two large, slow moving competitors) with an innovative strategy.

Breakthrough! Innovation Management in Practice

Positive, visionary outlook

An innovative executive looks forward to the future with a sense of optimism and a vision of success in development of new technologies, new processes, new products, and solutions to tough problems.

Impatience with business as usual

The innovative leader is continuously challenging the status quo as less than an optimum condition and promoting the belief that solving today's problems and meeting new challenges will inevitably change the organization for the better. This leader also conveys the notion that being overly risk-averse will lead to decay and the loss of market position and key customers. Innovators are concerned about the risks and costs of standing still.

Commitment to an innovative culture

The manager of an innovation project gives considerable thought and energy to developing and maintaining a good climate—where employees are inspired to generate and experiment with new ideas, including those which may not appear feasible. This leads to a form of disciplined patience in regard to development and implementation of new ideas.

Tendency to set ambitious targets

In addition to challenging the status quo, the innovative leader sets new and highly ambitious targets for themselves and their subordinates. Ambitious (and perhaps unreasonable) targets often lead employees to discard conventional ways of thinking about the business. In many cases, this is precisely what's needed to accomplish disruptive changes and implement new-to-the-world product ideas.

The innovative leader also clarifies her/his awareness that innovation projects, at various stages in the process (e.g., in Figures 3-1 to 3-4), typically involve several stages of diligent experimentation, failure, learning from mistakes, designing the next experiment, etc.

Exploratory versus judgmental thinking

In a culture where ideas are quickly evaluated and judged on their risk characteristics (what might go wrong), innovative strengths cannot be fully developed or applied. Innovative managers have an ability to consider new ideas from an exploratory viewpoint.

They tend to look at the positive attributes of new ideas and ask: "How can we experiment with this idea to understand its potential value"?

Relationships of trust

Another prevalent characteristic of the innovative leader is a high level of trust in the creative ability and judgment of subordinates. If the manager believes in the theory (and reality!) that everyone has creative talent, this will be a key factor in maintaining a good climate for innovation. Employees will be more willing and motivated to offer fresh thinking about problems, challenges, and opportunities—even when they are not sure about the benefits and costs of their embryonic ideas.

In a truly innovative culture, employees shouldn't be overly concerned about feasibility, attractiveness or risks of new ideas. They have an understanding that not all ideas will be carried through to implementation. They derive much of their reward from being actively involved in the teamwork, shared creative spirit and process of innovation.

Detecting the Need for Innovation

In establishing goals for the future, there are opportunities for a company to improve its competitive position and increase financial strengths through innovation. In general, there are two major categories of vital signs that drive the need for innovation programs:

- *In the key operating activities of the business* (research, product development, purchasing, plant scheduling, production, etc.), are there major problems or weaknesses relative to the competitors? These problems are usually obvious signs that innovation is needed.

- *In the dimensions of performance valued by customers* (price, quality, safety, performance, service), where are the opportunities to improve the company's position relative to competitors' offerings? These opportunities may signify the need to invest in innovation.

This does not suggest you should always use innovation strengths in efforts to beat the competition at its own game. Instead, creativity and innovation may be used to discover a niche where competitors are avoided altogether. Companies have succeeded with this strategy by finding new ways (e.g., innovative products and services, attractive prices, specialized ad networks) to deliver value to unserved or underserved markets.

Building Innovation Management Strengths

Management can obtain leverage from the organization's innovative capacity through its decisions and actions regarding:

Process

Ensure the organization has an innovation process and climate that enables generation and nurturing of new ideas.

Creating and maintaining a healthy climate is one of the most important activities in managing the process of innovation.

Communications

Build and maintain communication strengths! Promote open, non-threatening and exploratory discussion of new ideas and concepts, even if some of these emerge from external sources.

Rewards

Find ways to motivate and reward managers and employees who exhibit attitudes and behavior conducive to innovation.

Portfolio

Develop and maintain an *idea portfolio*. The portfolio owner should promote a continuous flow of ideas addressing a wide range of issues.

The portfolio serves as a springboard when fresh thinking is required.

Focus

Determine what areas of the business have the greatest need for change and focus the process on these areas.

Generate new ideas (or select ideas from the idea portfolio), experiment with a few ideas, and begin taking the innovative steps that yield desired changes.

5 | Communications and Trust

Communications and trust between the members of an innovation team and between the team and external stakeholders are perhaps the most neglected aspects of innovation management.

My observations reveal that success in innovation projects is dependent on high levels of trust and quality of communications. This is especially true when large numbers of people are involved in the process of innovation, and they must communicate across language, cultural and geographic barriers.

As critical success factors, these two elements of innovation often determine how and if a new idea or concept is implemented. And these elements of innovation are frequently not discussed in planning stages or mentioned in early stage project documents. Managers simply assume the team will have good communications and sufficient levels of trust.

If there is a low level of trust and poor communications between team members, this may not be visible or understandable (at least in early stages of development) to managers and team leaders. It could be a problem that even the affected team members will not be completely aware of nor will they want it to surface, because everyone is encouraged in the early stages by the energy and optimism of the opportunity, and few want to damage this momentum.

But the reality is that nearly all innovation teams experience various issues concerning trust and communications. These issues need to be addressed in all stages of the process. This chapter discusses the nature of such issues and provides recommendations on how to deal with them.

Innovation Projects and Knowledge-Sharing Networks

Although a specific department or group often drives development of new products and services, many ideas are discovered and concepts developed through the work of cross-functional project teams. For management, the project form of organization offers a number of advantages for innovation. The project form is a responsive, flexible and fast-moving entity—one that allows for communication and collaboration of informal networks of experts and partners. A project can be easily established and is easier to shut down than a department. (Although shutting down may get progressively more difficult as the project progresses to later phases and takes on new partners.)

Collaboration, knowledge sharing and shared risk-taking are all part of innovation projects and teamwork. Therefore, team members must have high quality of communications—mostly informal among themselves and perhaps more formal with external stakeholders. As the team forms and the project enters its initial phase, email addresses and mobile phone numbers are exchanged. Within a short time, an *innovators network* and knowledge-sharing communications are established among team members and between team members and other subject matter experts.

Collaborative software tools may be used to enable team members to view each other's work and share information regarding schedules, meetings, and other administrative matters. If team members are dispersed over multiple geographies, then the network also becomes their workspace and perhaps their meeting space. Information is exchanged through the network in a series of transactions.

Problems with Information Disconnects

Based on the work of R. S. Burt (1992), such networks often suffer from missing links, and this results in a gap or hole in the flow of information. A *structural hole* between two organizations or network participants exists if the parties do not exchange information or communicate adequately. The implication of Burt's work is that the innovation team needs someone to facilitate and improve communications among disconnected participants.

Why do some members of a network become disconnected? The reasons vary from one situation to another, but include the following:

- Barriers and rules that prevent communications with external parties
- Organizational rules on what can be communicated to external entities (and through which channels)
- Language or cultural barriers
- Lack of trust due to situational risks or other issues
- Technology issues
- Time zone differences
- Other social interaction issues

Breakthrough! Innovation Management in Practice

As we have seen in the process of innovation, the disconnection of participants (and breakdown of innovation transactions) causes problems in terms of idea development, prototype development, and implementation of new products and services.

Role of the Network Architect

When structural holes are discovered, this is another aspect of the innovator's challenge that demands someone assume a non-traditional role, for example the role of *network architect*. The person who serves in this role performs a wide range of activities critical to innovation success:

- Identifying the nature and cause of structural holes
- Clarifying to the project owner the impact of the holes
- Serving as broker or mediator to improve the flow of information
- Searching for ways to fill the holes on a more permanent basis

When innovation success depends on knowledge sharing or technology transfer across a functional or external border, and management becomes aware of a lack of communications or trust across this border, then the role of *network architect* becomes necessary. This role involves building an information bridge and collaborative partnerships that transcend the border. The parties involved may never be close friends or have long-term working relationships, but at least for the duration of the innovation project, they can communicate in a value-added manner and work together to achieve the objectives needed for success.

In addition to the organization's objectives for innovation, individual team members may have their personal objectives, which are not evident to others. Team members have a complex set of objectives and motivations, including the need for recognition among peers, the goal of getting promoted, or a strong belief in how a breakthrough solution might—as Steve Jobs memorably stated—"put a dent in the world".

Unfortunately, situations arise when a key functional department is not wholly in favor of an innovation team or its objectives. When this happens, the department's employees who are assigned to the team may not be highly motivated or supportive of other team members and may undermine the team's progress. In these situations, the network architect must work in concert with the project owner and facilitator to develop and implement problem-solving measures.

While formal power in most organizations is a function of position on the organization chart, informal power is usually determined by the scope and effectiveness of networking and other skills. Whereas the effectiveness of formal power depends on the ability to manage by accepted standards, processes and rules, informal power is acquired through the personal attributes of expertise, interpersonal skills and the trust to share knowledge and ideas across boundaries.

The individual with informal power usually has a high level of collaborative spirit. This person enjoys communicating with and helping people (regardless of their position in or outside the formal organization) and is an ideal choice for an innovation team. With the spectrum of advanced networking tools available, the number of networks and social / professional networkers is expected to increase, and it is up to decision makers to learn how their organizations can benefit from the flow of knowledge outside the organization chart. Following are recommendations on how to develop an awareness of informal communications:

- Develop an understanding of employee usage and trends in regard to social and informal networks

- Give people the opportunity to share ideas they gain through external networks (which allows them to demonstrate such networks help in reaching beyond the not-invented-here syndrome)

- Create and encourage the use of informal networks within the organization

- When innovation teams are formed, ensure that attention is given to understanding the team members' level of networking already in place

As described by W. Powell and S. Gradual (in Fagerberg et al., 2005) in their research on innovation networks, a "cluster of individuals that share a similar set of skills and expertise has been dubbed a community of practice (citing Wenger, 1998). . . .these loose groups are engaged in related work practices, though they do not necessarily work together. Such fluid groups are important to the circulation of ideas." Looking in more detail at the sharing of proprietary knowledge, Powell and Grodal noted (citing Saxenian, 1994) "that informal knowledge sharing, widely institutionalized as a professional practice in Silicon Valley, is one of the crucial factors contributing to its fertile innovative climate."

One way for a facilitator to gain awareness of an innovation team's informal communications strength (and weakness) is to create a *community-of-practice map*, as shown in Figure 5-1, where nine team members are depicted with the links indicating flow of communications and knowledge among the team and between team members and external sources of innovation.

In this example, several team members have informal links with experts inside and outside the team's boundaries, perhaps through their use of multiple social media channels. These team members keep the project owner and leader informed about conditions or changes in the formal organization or external environment.

Figure 5-1. Example of Informal Communication Patterns

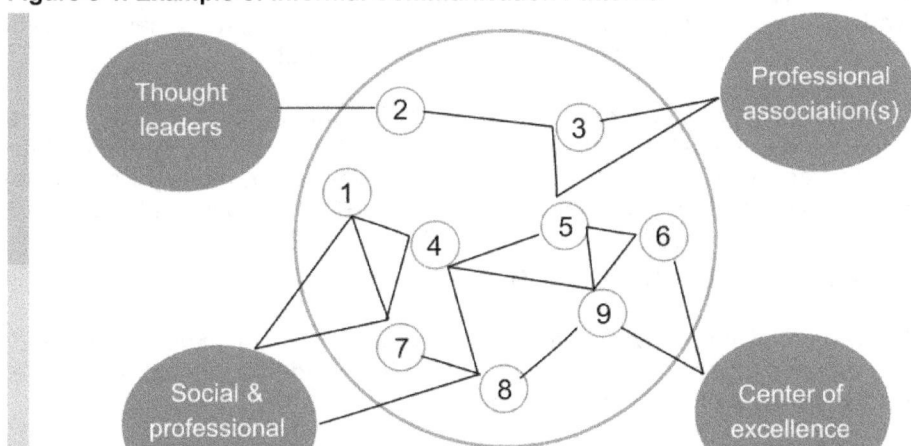

Innovation team and ecosystem

Although team member 2 appears to be disconnected from most other team members (only connected to member 3), it's probable that he or she is a valuable individual contributor through knowledge gained from thought leaders in the external environment. The facilitator should determine how to bring this contributor into the team's informal workings and dialogue.

With social networks evolving toward a mixed form of *social and professional networks* (SPNs), people in many fields rely as much or more on these forms of networking than on communications within the formal workplace. The current forms of SPNs enable people to network in new ways, and they help groups become more innovative. The challenge is how to tap into and capture the value of this growing body of knowledge.

Innovation Management and Trust

Success in managing innovation projects—whether the goal is introducing a new product, developing a new technology, implementing a new manufacturing process, or other challenge—is often dependent on whether the organization is successful in building a climate of trust. A good climate, with high levels of trust between managers and subordinates, between co-workers, and between the company and its stakeholders has a positive impact on building and maintaining innovation strengths.

Understanding what we mean by *trust* in this context and how it affects work relationships is vital, especially when we need to innovate. As many team-building activities have proven, trusting one's co-worker, manager or business partner is usually required before taking the next step into the unknown.

Understanding the Trust Equation

Between individuals and within groups, trust is a *function of knowledge sharing* plus *collaborative spirit* plus *role clarity* divided by the *complexity* (and risk) inherent in the equation:

$$\text{Trust} \approx f\left(\frac{\text{Knowledge sharing} + \text{Collaborative spirit} + \text{Role clarity}}{\text{Complexity}}\right)$$

Factors in the trust equation:

> ***Knowledge sharing*** . . . *the skills and willingness to share expertise acquired through experience, knowledge, specialized training and management coaching*

> ***Collaborative spirit*** . . . *energy and excitement about cooperating and communicating that enables a team to address a complex situation*

Role clarity . . . *the ability in new-to-the-world situations to define roles and have team members adopt non-traditional roles for the sake of experimentation and risk-taking*

Complexity . . . *a condition that results from the nature of the problem, risks, and the range and diversity of various outcomes (some of which are unfavorable to the decision-maker)*

Although the factors in the trust equation are not quantifiable, the value of the expression lies in recognizing that complexity inherent in innovation projects is mitigated through attention to knowledge sharing, collaboration and role clarity. If complexity is high, then ensuring a climate of trust demands that knowledge sharing, collaborative spirit, and role clarity are maintained at high levels.[*]

Trust and Decision-Making Roles

In many businesses, a few key decisions made by an individual, management committee or project team can have a major impact on success or failure of a promising innovation. In some cases, failure to make key decisions in a timely manner results in decreasing the value of the innovation. Trust is a key ingredient in decision-making situations, especially when the situation is complex. The complexity on an innovation project has many dimensions and may involve financial implications, product liabilities, intellectual property threats, safety challenges, environmental hazards and career risks.

A characteristic of today's business world is that most companies face a wide range of high-risk conditions. As unpredictable situations arise, strains are placed on existing decision processes, decision-making roles and relationships—because these evolved to meet the needs of a more stable and predictable state of affairs that no longer exists.

Making an Investment in Higher Levels of Trust

Making an investment in trust levels—through training, coaching and delegation of decision-making tasks—should yield a return in terms of understanding decision roles, streamlining processes and developing leadership strengths. These are key ingredients for successful innovation.

---

[*] For another perspective on trust—with a suggested trust formula (from Synectics® methods)—refer to Koppett (2002), Chapter 2.

Suggested actions to increase trust levels, improve working relationships, and build innovation management strengths:

Examine trust levels and determine where changes are needed to improve knowledge sharing, build a spirit of collaboration, and reduce complexity and risk

Evaluate existing processes and identify where complexity or rigidity hinders innovation. (For example, do employees devote too much time and energy avoiding complexity, rather than attempting to find and implement new, value-driven approaches?)

Give employees the opportunity and encouragement to modify and improve processes (while removing outdated barriers to innovation and minimizing the risks of presenting novel approaches).

Maintain awareness that the trust equation is in the background of nearly all change management initiatives and decision-making situations. If we maintain this awareness, it helps in streamlining the process for key decisions by enabling us to manage the factors of collaboration, role clarity, knowledge sharing and complexity.

On selected projects with multiple decision points, develop an understanding of roles and responsibilities. Ensure that opportunities to explore potential new products, services, and processes are identified, and encourage employees and business partners to present ideas and new concepts, even when these are not clearly articulated. (Good ideas often originate with rough edges!) This provides opportunities for people to exercise and test their skills in development of new ideas and breakthrough solutions.

Recognize in today's era of networking—when people rely on many types of professional and social networks—that external relationships are crucial in the process of developing innovative energy. The mindset to trust information from external sources is now an innovation success factor. (Chesbrough, 2003)

As discussed in Fagenberg et al. (2005),"cultivating the capacity for absorbing outside knowledge, so-called *absorptive capacity* (Cohen and Levinthal, 1990)" is something that "firms often find very challenging; the *not invented here* syndrome is a well-known feature in firms of all sizes."

6 | Innovation and Persuasion

To energize an innovation project, the most critical times are often at the point when the team must make a transition from idea generation activities to project approval and funding. If decision makers are risk averse and entrenched in the status quo, then the task of persuading them to accept or fund a proposed innovation is an uphill climb. This is when skills in persuasive communications add essential value to the process of innovation.

Defining Persuasive Communication

Persuasive communication = the process and practices needed to develop and present messages, models and analyses for an audience that the writer wishes to influence

Examples of persuasive communication include:

- Business plans and strategy documents
- Funding applications
- Customer presentations
- Marketing brochures
- Website and social media content
- Thought leadership, white papers and position papers
- Proposals and bid documents

Persuasive versus Technical Communication

Although technical communication is a fundamental part of many large projects, it has its own set of standards and best practices—which differ from those employed in persuasive communication methods. In general, technical communication is directed at making complex information and data readable for a target audience.

However, persuasion focuses on the small amount of information that influences decision makers and opinion leaders. Table 6-1 summarizes the differences between technical and persuasive communication.

Table 6-1. Technical versus Persuasive Communication

Technical communication	Persuasive communication
Intellectual response	Balance of emotional & intellectual response
Impersonal & objective	Often personal & subjective
Emphasizes features	Emphasizes benefits
Information driven	Influence driven
Structure is logical	Structure is situation dependent
Designed to make complex data & information readable	Designed for specific decisions

Recognize Audience Characteristics

The persuasive communicator recognizes that a background of knowledge, experience, and attitude toward risk and complexity will affect a decision maker's opinion about a proposed innovation:

- All decision makers and decision-making bodies have a *set of standards* that guide their behavior and thinking.

- Most decision makers have a *mental model or scenario* of the environment in which the organization operates (e.g., macroeconomics, markets, competitors, technology trends, regulatory issues, etc.).

- Each decision maker has a *threshold of acceptability*, based on the mental model, personal standards, beliefs, and intuitive judgment on opportunities, risks and success factors.

Apply Persuasive Writing Skills

Developing and applying persuasive writing skills is not simple. As stated by Bettinghaus and Cody (1987): "It is sometimes easier to say that one should get the intended message across than it is to create messages that actually ensure the message gets across."

Following these basic practices will help in getting your persuasive message across:

- Understand decision makers and audience characteristics
- Clarify desired results and define your strategy
- Identify key persuasive messages (don't overwhelm with data)
- Focus on perceptions
- Create a structure that suits the audience expectations and positions your key messages
- Build trust!

Maintain Awareness of Audience Expectations

A success factor for persuasive communications and building trust is your relationship with the audience. If decision makers in the audience have an entrenched mental model that conflicts with the proposed innovation, then you must ask some key questions: Should we expect the audience to *accept* a proposed innovation (which may involve a large commitment of resources)? Or should we have a more realistic goal of persuading decision makers to *listen* and perhaps approve a small step toward building a model—which involves a smaller commitment and less risk?

When a decision maker feels she must change her view on matters involving her strongest beliefs, then she may resent that she needs to compromise her own beliefs and vision in order to accept your view and commit to a large allocation of resources.

If your goal—in terms of audience expectation—is realistic, the chance of influencing the decision maker is greatly enhanced. As a starting point, consider writing an *idea profile*—which includes a brief description on the nature of the idea, its positive features, areas of concern, and proposed next steps. The idea profile serves as a tool to clarify the idea and build momentum toward acceptance.

**Anatomy of a
Persuasive
Idea Profile**

An "idea profile" is a written tool used to clarify the ideas, create a dialogue and build momentum. The challenge for the innovator is to ensure the reader's interest is awakened and maintained.

If the idea is similar to something tried before, then the innovator must demonstrate why this version of the idea has merit, offers potential benefits, and should be further developed. The innovator should not be concerned with a benefit / cost analysis at this point.

If the idea is new-to-the-world, then the innovator should include the idea profile as part of an imagination-stretching (breakthrough) scenario.

Craft Scenarios to Manage Complexity

In most business and government planning, the situation that management encounters is complex and volatile. Scenario planning offers methods for dealing with situations characterized by complexity and uncertainties.

It is natural for people to view the future as a gradual evolution of the status quo. This natural tendency makes it difficult to consider a future outcome that is very different from today. This is why many strategy development activities focus more on small changes in the status quo and fail to consider scenarios with far-reaching opportunities and consequences.

An organization may have a strategy of moving toward a narrow range of *accepted scenarios with similar features,* but by evaluating a wide range of potential scenarios, they become more prepared to watch for signs of disruptive change that signal the need for strategic redirection. Scenario planning gives the organization a way to visualize a wider range of market, competitive and technology situations.

For the strategic planner, scenario planning consolidates large amounts of information and enables the planning group to focus on a manageable number of possible outcomes. Each scenario tells a story. Various elements and uncertainties interact under different assumptions. Ultimately, each scenario becomes a credible description of a conceivable future.

For the innovation manager, scenario planning provides an opportunity to *challenge the status quo* and current management beliefs.

The difficulty occurs when the innovation manager believes in the plausibility of a scenario that suggests a radically different future driven by one or more disruptive forces, and the organization's leadership does not agree or is unwilling to contemplate the scenario.

In the late 1970s, computer companies worldwide were on the verge of massive changes in their industry. These changes involved the introduction and rapid acceptance of personal computer technologies. In their strategy development process, some computer companies did not contemplate or prepare for a *PC scenario* and eventually disappeared from the scene.

In another example of narrowly defined scenarios (circa late 1980s), telephone-operating companies were on the edge of the broadband and online services revolution, but would not include such scenarios (which were considered not plausible) in their planning process.

In describing a scenario with massive changes and a paradigm shift, the innovator must develop a compelling story.

Seven-step Method to Write a Compelling Scenario

1 · Define the issues that require management consideration

2 · Identify forces acting on the organization

3 · Clarify industry and market trends that affect the issues

4 · Understand the stakeholder audience and their preconceived views regarding the future

5 · Identify major uncertainties and possible outcomes and how these affect the issues and trends

6 · Create a matrix of scenarios and show how a few shifts in demand, competitors' moves, or emerging technologies could re-shape one of the scenarios, transforming it into a radically different situation

7 · Use the scenario with industry-changing features to *stretch toward a new paradigm involving a breakthrough innovation*

Breakthrough! Innovation Management in Practice

Working in the background of the scenario-writing process, an innovation team can contribute important knowledge and perspective by creating a Camelot scenario, as discussed by Higgins (1994) and other practitioners.

With this technique, participants are asked to articulate their version of a "Camelot" situation (the ideal solution), compare this with the status quo, and identify the innovation (and breakthroughs) needed to leapfrog barriers to arrive at Camelot.

Integrate Innovation and Thought Leadership

In some situations, organizations decide to share their ideas and knowledge in the marketplace. The objective of this knowledge sharing, in the context of persuasive communications (and *open innovation*), is to demonstrate thought leadership on key challenges and issues of concern to customers, policy makers, investors and other opinion leaders.

> *Thought leadership* = the intersection of knowledge management and marketing communications to demonstrate leadership in topics of interest to customers, partners, suppliers and other stakeholders.

A thought leadership program may have one of many possible forms, from a market research study with analysis of consumer behavior to a white paper with recommendations on how to deploy an innovative technology platform.

Information developed from thought leadership is made available to a target audience through proposals, business plans, published reports, articles, white papers, trade shows, website content, webinars, podcasts, and social and professional media networks.

Investing in a structured thought leadership program should yield benefits such as:

- Establishing a reputation as a leading innovator
- Appealing to potential business partners and other stakeholders
- Attracting investment, venture capital funds, and other financing
- Influencing policy makers and regulatory authorities
- Emphasizing a culture of innovation and attracting prospective employees
- Setting the stage for launch of new technologies, products and services

Thought Leadership Opportunity in the Passenger Rail Sector

On a project involving refurbishment of rolling stock, a leading rail manufacturer and its competitors wrote bid documents and made presentations to the technical staff of a public transit authority in a major US city.

For the manufacturer in question, it became evident to the management and engineering staff that for one of the vehicle subsystems most crucial to the transit authority—involving the vehicle monitoring and control system (VMCS)—the authority preferred the one proposed by this manufacturer.

Although the latter company eventually pulled out of the bidding on this project, there should have been internal recognition that the company was indeed ahead of its competitors in terms of VMCS knowledge and innovation. This technology edge presented the possibility for thought leadership and technology innovation, but decision makers overlooked the opportunity.

Rely on an Editor to Make It Convincing!

Most executives and managers who have been involved in developing business documents to persuade an audience or decision maker—such as bid documents, strategy documents, business plans, white papers, and product development proposals—are aware that the process of planning, writing and producing such documents is complex. This is especially true when the research and writing tasks are distributed among the members of a large project team, and the team members are not professional writers.

Adding to the complexity of the writing process are the myriad issues that result from different views on document objectives and styles; internal politics; debates surrounding technologies, strategies, markets and competitive trends; numerous departments involved in reviews and revisions; and other unpredictable problems. These hurdles add to the difficulty of writing a persuasive document and may cause the team to lose sight of the audience.

To manage document complexities—while maintaining awareness of the audience and achieving *high-impact persuasiveness*—most large projects need an experienced editor. In fact, the success of the document in having the desired impact on the target audience often depends on the skills of a professional editor who, in the role of collaborator, facilitator, coach, and perhaps ghostwriter, improves the writing and adds value to the work. The editor becomes an active partner in the process of creating conversation and dialogue.

An editor helps the project team and writers in planning the document; crafting early-stage versions (creativity is often the focus at this stage); and making revisions that ensure a convincing and credible document. Therefore, the role of the editor is to deliver—in addition to language skills—a unique mix of creative, technical, strategic, facilitative, and diplomatic skills.

Learn the Art of Innovative Conversation

It is a curious thing about creativity and innovation that as we venture into uncharted territory, we search for the right words and phrases that help us describe a vision of something new, perhaps something that no one else has seen or thought of in the way we are thinking about it.

Although we are convinced that our idea or strategy will work, we struggle to describe it in words. Maybe we go to the white board and try to create a sketch or picture of what we see in our mind. If we convince a few others that our idea holds promise, then we may keep it alive for a period of time.

Facilitate an Innovative Conversation

As we move an idea toward concept development, others form their own version of how it might work. During this time, new words, phrases and acronyms will arise and become part of the project vocabulary and dialogue. The project team may not be aware that a unique *language or dialect* has emerged.

"The language spoken by innovators sounds like a foreign language. It is unique, often colorful, and relies on acronyms and phrases from research, development, academia, emerging business and sociotechnology trends . . . and includes terminology from the two modes of thought known as *analytical, left-brain thinking* and *intuitive, right-brain thinking*. In a successful innovation project, these modes are complementary" (Glasco, 2013).

It is imperative in the early stages of the innovation process that team members are aware of the importance of the exploratory mode of thinking and communicating, a mode where it's permitted—*and encouraged*—to try different ways of expressing new ideas. And it's also permitted in the early stages—and expected—to make mistakes!

Breakthrough! Innovation Management in Practice

If an emerging idea or concept is similar to something that's already been tried without success, then it may also be necessary to position the concept in a different way, with different terms and acronyms, and to clarify how the concept has been modified from the previous versions. This is part of the challenge of developing an innovative dialogue.

Clarify the Intended Message

One of the complex aspects of describing a new idea or concept in a meeting or other interpersonal situation is that the idea generator and the audience often discuss something that neither of them has ever seen or considered. By necessity, there will be new language and concepts that must be discussed and framed in an evolving context.

This presents the problem of the audience not hearing the same message or seeing the same picture in their mind that the idea generator is attempting to convey. As explained by Nolan (1987), this is known as the *gap between intent and effect*. Going further, Nolan states that "correct hearing and understanding of the content of the message is not enough for full communication. There also needs to be a match between my purpose in telling that person (my 'intent') and the effect the message has on him or her."

Therefore, the facilitator must help by encouraging the listener to paraphrase what was heard from the idea generator. When this is done in a positive way—e.g., *"Here is what I understand to be your idea."*—it sends a message that the listener is making a genuine effort to understand the idea (rather than evaluating or attacking it). The paraphrasing technique clarifies what is intended by the idea generator, and improves the chances that the idea will stay alive and gain support.

A Few More Tips on Persuasive Communication Methods

Use headlines to emphasize the key points

If analytics are needed, use only the summary (save the details for later)

Summarize the risks of making the decision versus not making it

When appropriate, show all sides of the issues

Include a vision of success and clarify next steps

Be realistic in estimating resource requirements for the next steps

Breakthrough! Innovation Management in Practice

Breakthrough! Innovation Management in Practice

7 | Non-Traditional Roles in Innovation

"If you deeply appreciate and love what creative people do and how they think, which is usually in unpredictable and irrational ways, then you can start to understand them."

Bernard Arnault

Creative and innovative people do not always fit comfortably in traditional organization charts and one of many challenges of an innovative leader—when confronted by an unexpected situation or opportunity outside the mainstream of the business—is to encourage selected individuals to adopt non-traditional roles.

Most members of a project team or task force, when assigned to develop something new, must initially represent a department or functional area. However, disruptive situations demanding a new approach may not conform to the current mode of operation or organizational schema. Innovation projects demand leaders to break through the lines on the organization chart and encourage (and reward) people to think, behave and work in new ways.

Consider these examples of creative people assuming new roles and responsibilities during innovation projects:

- An engineer accustomed to designing hardware proves to be talented in generating fresh ideas for end user solutions.
- An operations manager embraces the role of sponsor for several promising ideas and earns a reputation for persuading others to fund and support the ideas.
- A legal and regulatory analyst proves to be skilled in building trust among team members and with external stakeholders.
- A marketing manager views intellectual assets in relation to weak signals from the market and produces a breakthrough product concept.

Looking beyond the traditional roles assumed by content experts (who are the main idea generators), Table 7-1 describes some of the roles important to the innovation process.

Table 7-1. Roles and Responsibilities on Innovation Projects

Innovation Roles	Role Characteristics
Owner	Has highest degree of ownership in the innovation; provides *content direction*. Effective motivator
Experimenter	Skilled at generating ideas and finding low-risk ways to experiment with the ideas and develop them further. Looks at ideas as the clay needed to form a prototype
Facilitator	Has minimal content involvement; provides process direction. Guides team members through the process and helps define their roles
Idea champion	Promotes new ideas, finds ways to nurture and develop ideas, advises the Owner and Project Leader on content value of new ideas
Sponsor	Usually a good motivator who provides initial impetus for the innovation and defines the desired outcome
Visionary	Provides conceptual views and ideas on future opportunities. Adept at describing future scenarios and sketching a *breakthrough scenario*. Demonstrates entrepreneurial spirit
Diverse Resource	An idea generator recruited from the project periphery; adds new perspective and fresh ideas
Communicator	Provides expertise in documenting and articulating plans and results; skilled in writing persuasive communications and crafting descriptions of new solutions, concepts and breakthroughs
Network Broker or Architect	Maintains awareness of key contacts and resources in adjacent networks of interest to the project; assists the team in drawing knowledge from other networks
Project Leader	Manages the project on behalf of the Owner; often shares the role of content direction with the Owner
Scout	Provides information on related events, opportunities and issues within and outside the organization
Implementer	A tactician capable of turning ideas and concepts into reality. Advises the Project Leader on implementation methods and challenges and is a good problem-solver

The Role of Facilitator in the Innovation Process

Much has been written about the role of the facilitator in meetings, with emphasis on guiding the process of meetings (and not getting too involved in the content). In addition to this role in meetings of all types, the facilitator role is essential in designing and managing the process of innovation. While ideas may emerge in meetings, it is outside the meeting room where the team must roll up their sleeves and get on with the task of developing, nurturing and implementing the ideas.

Many new ideas emerge during the course of performing routine work on the project. The facilitator is a specialist to whom the project leader and team members can call upon when such ideas emerge—to ensure ideas are documented in some manner and included in the innovation dialogue.

Regrettably, it's a fact of corporate life that after the enthusiasm generated during brainstorming meetings, the team loses momentum. An idea that initially looked attractive starts to look tarnished and less interesting. Other urgent projects take priority. A leader who at first appeared to be a committed sponsor for the idea is suddenly less visible or perhaps changes jobs. It is during these periods of decreasing momentum that the facilitator must take action to get the team back on track.

Role of the facilitator

The role of facilitator varies significantly from one project to another. In many situations the facilitator serves as a process specialist (i.e. does not get involved in the content of the project) who plans and manages team meetings and idea generation sessions.

In more complex environments, the facilitator *guides the day-to-day process of the innovation enterprise,* while also serving as negotiator, motivator and network broker.

Another responsibility for the facilitator involves problem-solving activities. Most innovation project teams and the technical experts on them are frequently in a problem-solving mode. Team members will have divergent views on how to solve problems. They will compete for resources. Discussions about potential options will take place—in text messages, in the cafeteria, during informal meetings, and in local watering holes—but may not lead anywhere in terms of deciding on next steps.

Therefore, a key activity in the project stalls. Momentum is lost. The project schedule and just as ominously the nature of the project outcome are at risk. This is a time for the facilitator to step forward. Problem-solving activities can take place in meetings with a subset of the project team, or in some cases the entire team needs to be present. One of the techniques that the facilitator must employ effectively is pre-meeting planning. This will typically consist of:

- One-on-one interviews with stakeholders to clarify problems
- Requesting team members to submit written summaries on the nature of the problems and proposed solutions
- Identification of a leader who has a vested interest in restoring momentum. This will be someone who serves as the sponsor in a problem-solving meeting.
- Selecting a location for the meeting that is on neutral ground and is conducive to collaboration
- Deciding on the meeting attendees. Generally this needs to be as small as possible (less than ten people)
- Coaching the attendees on the objectives of the meeting and their roles. This is especially valuable for diverse resources who are not directly involved in the problem itself but provide expertise and an unemotional viewpoint

Types of Meetings

There are many types of project meetings where the facilitator adds value:

- Project planning meetings
- Ideation and brainstorming sessions
- Problem-solving workshops
- Idea development meetings
- Strategy workshops
- Proposal development meetings
- Innovation audits

Some types of project meetings and workshops (especially those involving innovation, problem-solving and change measures) should be relatively unstructured, and this is where the role of facilitator role is vital to success.

Structured meetings	Less structured events and interactions
Staff meetings	Strategy workshops
Project planning	Brainstorming meetings
Project status	Problem solving
Milestone reviews	Idea development
Presentations	NTW* concept planning
Sales meetings	Design meetings
Interviews	Innovation audits

* New-to-the-world

A Visionary Looks Beyond Existing Boundaries

If your organization requires a breakthrough solution or a creative response to a disruptive situation or threat, then the role of visionary is indispensable. This role extends beyond individual projects. The visionary is someone receptive to new ways of thinking and capable of articulating an emerging concept, business model or solution in persuasive language that inspires and motivates innovators and entrepreneurs.

In developing alternative solutions for a complex challenge, the innovation management issue for many organizations is an inclination to lean toward a vision considered feasible in terms of *current* technologies, products, and processes. But feasibility is sometimes an evolving concept, subject to market and competitive forces that are impossible to predict. This is the impetus for maintaining an *innovation portfolio* and for motivating managers, employees and business partners to explore beyond the boundaries of current feasibility when pursuing breakthrough solutions.

The Nature of Innovators: Risks and Rough Edges

A colleague of mine once described creative people as self-reliant individualists with lots of *rough edges* (e.g., radical ideas, unusual work habits, and an enduring belief in the value of their creative skills). He claimed that it was an unfortunate aspect of the corporate world that large organizations with established cultures had an unavoidable propensity to gradually suppress or *grind down* the rough edges of new employees.

71

The objective (usually unstated and unintentional) in these situations is to reach a point when the new employee has no rough edges—to achieve an effect like smooth pebbles at the bottom of the stream. Everyone looks alike, thinks alike, avoids risks, and allows the water to glide smoothly past without making a ripple. However, the new employee's rough edges offer potential value in terms of discovering knowledge and ideas from external sources and finding unique ways to look at innovation challenges. The experience and knowledge gained in a previous organization enables the new employee to give fresh insight on an opportunity, problem or barrier to innovation.

Capture the Value of Rough Edges

In some innovation projects, a facilitator may discover that a team member who appears isolated is a relatively new employee and an independent thinker. Unfortunately, most formal organizations require the new employee to undergo a process of employee orientation. The new employee learns how things are done in the organization. Rules, procedures and standards of expected behavior must be followed.

And the new employee is—at least indirectly—encouraged to unlearn or suppress old behaviors (and in some cases experience and knowledge!) from previous jobs where he or she may have worked. In other words the formal organization looks at the new employee as someone with rough edges and starts the process of smoothing those edges so the employee acquires the behavioral and rule-based characteristics of co-workers.

The facilitator or project leader should take time to meet with the new employee to discover what prior knowledge he brings to the effort, encourage him or her *to innovate from the outside*, and contribute a perspective and ideas based on prior experience.

Rough-Edge Innovators

In my work with Fortune 500 companies and multinational partnerships, I have observed three major types of innovative thinkers and doers. I call these rough-edge individuals the *Tethered Explorers, Adventurers* and *Extraventurists.* They each have rough-edge ideas and thoughts from time to time. They tend to challenge the status quo. One of them remains close to and is more dependent on the organization, and—although he or she creates outside the box—will return to it to implement ideas and concepts. The other two innovators are less averse to the risks and—assuming they have enough independence—will not only generate ideas outside the box but will also implement their ideas outside the organization if given an entrepreneurial opportunity (or if their ideas are not accepted by the organization).

72

Tethered Explorer

Some people like to test the boundaries of their organization's culture and risk-taking attitude by proposing ideas that simply don't fit in the box. They are good at generating a new product idea or business concept and experimenting with it in some way. But whatever they do, they do it with the quasi-approval of their organization. Like the space-walking astronaut who is tethered to a spacecraft and relies on it for life support, they remain connected to the organization and will eventually bring their results back into the organization. These innovators are the *Tethered Explorers*. They are important to the organization but usually are not big risk takers on their own.

Adventurer

Some innovations will simply not find a *strategic fit* in the organization chart. However, there might be a sponsor who believes in the value of a new idea and convinces senior management to invest in the project— with an agreement that it is implemented outside the chart.

If the project is successful, then it may be established as a subsidiary and eventually be folded back into the organization. Innovators who start these types of projects are aware of the challenges and associated risks. They are *Adventurers,* but usually need the support of the organization to get their ideas off the ground.

Extraventurist

This is an aggressive innovator with an entrepreneurial mindset and plenty of rough edges who resists virtually all attempts at organizational smoothing.

This is an innovator and visionary who might not trust the organization with his or her ideas—and may decide a rigid culture is not conducive to breakthrough ideas. If the *Extraventurist* is not recognized or encouraged, he or she is reluctant to share the most transformative ideas, and will eventually search for ways to independently form a startup enterprise.

Rough-edge thinkers offer value to an innovation team in terms of their entrepreneurial ability to generate and develop breakthrough ideas and new-to-the-world concepts.

Breakthrough! Innovation Management in Practice

8 | Barriers to Innovation

"Difficulties are just things to overcome, after all."

Ernest Shackleton

Although ideas for innovation projects usually begin with an initial surge of excitement and energy, the harsh reality is that many innovation projects never get off the ground—or get terminated during early stages of development—due to insurmountable barriers. Therefore, a critical step toward success of a promising innovation is to recognize and confront the barriers.

The barriers to innovation are numerous and include:

- Funding shortfalls and insufficient investment
- Lack of innovation management knowledge and skills
- Insufficient access to sources of research and technology
- Lack of vision and low awareness of commercial opportunities
- Regulatory and legal complexities; intellectual property issues
- Poor communications with partners and other stakeholders
- Poor internal climate for innovation
- Aversion to perceived risks
- Not-invented-here mindset
- Lack of appropriate government incentives or policies

Table 8-1 provides more examples of barriers at various levels of innovation.

Table 8-1. Barriers to Change and Innovation

Innovation Level	Barriers
MIP (multinational innovation programs)	Poor networking and low trust levels
	Political and cross-border disputes
	Lack of leadership
	Financing shortfalls
NIS (national innovation systems)	Poor networking among key economic clusters
	Inability to focus on high leverage projects
	Lack of high-impact links among government, academia and private sector goals
	Threats to established power relationships
LIP (large infrastructure projects)	Resistance from special interest groups
	Inability to overcome funding challenges
	Disagreements on benefit-to-cost ratio
	Threat to environment or cultural sites
	Lack of persuasive communications
TPS (technology, product & service innovation)	Individual and group conflicts
	Lack of vision concerning opportunities
	Lack of process management skills
	Technology or implementation complexity
PSI (problem-solving initiatives)	Poor climate for change
	Lack of ownership
	Lack of persuasive communications
IIP (incremental improvement projects)	Poor climate for change
	Entrenched process
	Perception of low value

Culture and Climate Barriers

Although many executives emphasize the importance of innovation in their companies and industries, they are sometimes not aware that a poor corporate culture—the internal *climate for innovation*—is a barrier to innovation.

Occasionally, new ideas and proposed changes are presented in a poor climate, but if decision makers favor the status quo, then undue pressure is placed on the innovator to prove the value of an idea before it is fully developed and before analytics are available. There is seldom a chance to resubmit the idea after management rejection (even though management may not intend for the idea to die). To detect a poor internal climate for innovation, be aware of these symptoms:

- Aggressive questions in early stages about risks, costs and implementation (suggesting that quantitative data is already available when usually it is not)

- No encouragement to explore the *nature of the idea*

- Lack of ability or willingness to consider the positive aspects of new ideas (*before* attacking the negative aspects)

- Few opportunities to apply persuasive communications

- Lack of an idea champion or sponsor at senior level

A successful innovator is by necessity a change agent who pushes, cajoles, persuades and emphasizes the need for change and breakthrough solutions. But many stakeholders oppose change for a variety of reasons (especially the changes involving a departure from the status quo or requiring large resource commitments).

An organization's culture—often influenced by a national culture which is highly sensitive to making mistakes—is also a poor climate for innovation. Another example of barriers is in organizations that adopt a "get it right the first time" attitude toward work. The process and work of innovation, by its nature, is *not* about getting it right the first time.

In the early stages of an innovation project, the sponsor and project leaders are facing an adversarial climate and culture when they observe some of the following signs:

- During meetings, key members of management appear to disengage and do not listen or participate.

- One or more executives take a dominant stance toward other team members and attempt to create competition within the team.

- Some team members are skeptical of the proposed changes and ask damaging questions or disagree with early-stage ideas and concepts.

- Someone insists that an established process (although it has its weakness) was mandated by management and is therefore unchangeable.

- Someone insists on achieving precision and high quality in the early stages of exploratory thinking and brainstorming (when precision and quality are not the objectives).

- A key stakeholder insists on having milestones defined in detail in the early stages of the process (when few details are available) before making a commitment.

- A team member is uncomfortable with and disagrees with the *unstructured style* of a brainstorming session (although this style is most conducive to producing fresh ideas).

For an innovative mindset to survive and flourish under difficult conditions, management must encourage exploratory thinking and be prepared to accept and learn from mistakes. This is part of the process of innovation.

Although the quality goal of "getting it right the first time" is worthwhile for work that is routine and part of established processes, innovation activities are generally not part of an established process. And a proposed innovation may ultimately lead to the creative destruction of an established process.

In many situations, a poor climate is the major culprit for innovation weakness and stems from a complex mix of factors.

Factors Contributing to a Poor Innovation Climate	Lack of management commitment
	Inability to break away from an ossified process or strategy; tendency toward Type II errors
	Rigid professional network with high barriers to entry (where participants become tightly knit and information flows only within a small group)
	Lack of communications and trust between the innovation team and stakeholders
	Risk averse culture, barriers to change, lack of exploratory thinking

Strengthening innovation management skills and knowledge enables an organization to develop and maintain an internal culture and climate conducive to innovation and change.

Institutional Barriers Constrain the Innovator

Although governments and policy makers will often extol the virtues and benefits of innovation in their business and academic environments, they may at the same time handicap the potential innovator through burdensome regulations, complicated tax systems, lack of protection for intellectual property, and legal constraints that make it difficult for firms to cooperate or accomplish technology transfer across national borders.

Regrettably, the innovator in such cases will spend excessive amounts of energy, time and resources trying to overcome the institutional barriers before getting on with the real work of innovating. Or the innovator may decide that a potential innovation simply does not yield a positive return in such an unfavorable climate, and will take his or her ideas and skills elsewhere.

Breaking Away from an Outdated Process

A constraint that serves as a frequent and difficult barrier to innovation is a *lack of process skills*. In particular, this refers to the difficulties in identifying and creating a process for innovation projects.

An organization may have a highly developed process for product development and other functional activities. But if a proposed innovation is a unique idea, perhaps suggesting a radical change, then an existing process may not suffice. This problem is often obscured in the early phases of an innovation project. If a new product idea emerges which extends beyond the current product strategy, people involved in new product development may lean toward an established process which is proven and understandable. Breaking away from the old process and reinventing it are often required for successful innovation.

When a new process is designed, the sponsor and facilitator must ensure that team members (and stakeholders) know at all times where the team is in the process. There may be times when the team moves into the concept development phase and discovers that a technology barrier damages their momentum. The project leader has to return briefly to the ideation phase to overcome such barriers. When this happens, team members must place the analytical aspects of the project on hold and return to the more intuitive and exploratory mode of thinking.

Preventing Damage to Early-Stage Ideas

In their earliest stage of creation, ideas are often presented in meetings. But meetings are seldom designed for the purpose of nurturing ideas. Problems

79

encountered in promoting products or systems, obtaining delivery of needed parts, attempting to reduce costs or increase quality, trying to achieve higher levels of customer satisfaction, or debating how to comply with new regulations may dominate the meeting agenda.

In some situations, employees are compelled to submit ideas. However, if the culture is accustomed to putting the innovator on the defensive or is resistant to change, then the new ideas—those that might produce a breakthrough—will seldom survive. Ideas are fragile things, and there are many ways to damage them, whether intentional or not.

Facilitator's Goal–Prevent the Top 10 Responses that Damage Emerging Ideas

Ignore the idea (the silent approach).

Declare that it's too costly. *"We could never get the funding approved."*

Claim that it would be too risky.

Point out that it would be too difficult to implement, and produce reasons why it won't work.

Mention that it has never been attempted.

Reveal that we tried something like this before, and it failed.

State that a competitor is already doing something like this.

Remind the team that it would conflict with other initiatives.

Make a personal attack on the idea generator.

Accept it but have a committee sit on it (or modify it out of existence).

After incurring damage, ideas may not reemerge in the near term, if ever. The innovation team needs to build a climate for innovation that prevents (or at least minimizes) harm to emerging ideas and concepts.

The facilitator can take several steps to prepare for potential opposition and damage to new ideas while building a healthy climate and overcoming barriers to innovation:

- Conduct one-on-one interviews to ensure that plans and objectives of idea generation meetings are understood by the participants

- Give participants and their stakeholders opportunities to voice concerns off-line from team meetings (giving assurance that concerns are recorded and *parked offline*)

- Coach team members on the process and language of innovation

- Ensure roles and responsibilities are clarified (for meetings and post-meeting actions)

- Use idea profiles (as discussed in Chapter 6) to facilitate communications and stimulate collective thought

- Ensure the innovation has uncensored opportunities to transform ideas into concepts (as discussed in Chapter 3)

The Need for Innovation Audits

Based on analysis of more than 1000 innovation projects, data scientist Thomas Thurston found that 78 percent of initiatives did not exist seven to ten years after receiving approval and funding. "Stated differently, three out of four of those initiatives either failed to reach scale or to become self-sustaining." (Hunt, 2015)

When innovation efforts fail, people associated with the project want to distance themselves from it. They want to move on to other work they feel is rewarding and offers the opportunity for personal growth. Management may ask for a summary of *lessons learned*—then place the results into an obscure file with the belief that "we'll avoid these mistakes in the future".

But the innovation project that fails offers valuable information that should benefit the next project. Most likely, the organization will not discontinue all innovation activities. How can an organization discover the nuggets and define the value from a failed project? One way is to conduct a thorough *innovation audit*.

This is especially constructive for large projects, where the lessons learned are scattered over a wide area with multiple departments, business partners, consultants, and other players holding key information and opinions regarding where and why problems occurred.

The innovation audit, led by a forward-thinking sponsor, should consist of at least four major components:

- Anonymous interviews with key managers and participants
- Workshops to *reverse engineer* the process and identify where barriers, conflicts or lack of trust impeded the team's progress
- Ideation sessions to create a new process model for use in future innovation projects
- Documentation that records the audit and makes the knowledge available to future project teams

In planning audit interviews and workshops, there should be an awareness that some project team members may not be as cooperative with each other as they were at the beginning of the project. There will sometimes be a breakdown of trust between team members or departments that believe original agreements and commitments were not upheld, deliverables were not provided, or other project milestones were not met—and therefore the other party caused the failure.

The innovation audit team must therefore tread carefully through this minefield of discontent and mistrust to find and articulate the high-value findings and knowledge. To conduct an in-depth innovation audit, it is advisable to recruit a new facilitator who will bring a fresh perspective to the process and an approach to *gaining knowledge* and presenting *positive results*.

9 | Innovation in the Public Sector

"Innovation in government is about finding new ways to impact the lives of citizens, and new approaches to activating them as partners to shape the future together."

OECD (2017)

Public sector organizations are often regarded (sometimes inaccurately) as conservative entities with little appetite for risk. While the mission statement of a government department may refer to its role as a provider of public services, the department's success as a service provider is not often linked to innovation or change, unless it has failed in a severe crisis and suffered criticism from the public and media, as in the Hurricane Katrina disaster (which involved the poor response and lack of coordination of emergency management groups, personnel and systems).

In contrast to the business world, the government leader does not—until recently—depend on innovation. In the business world, companies of all sizes and in nearly all industries depend on innovation at some time in their history. Innovation strengths and weaknesses affect a company's ability to create new products and services, take advantage of new technologies, and invent new processes. The private sector executive tries to innovate as a means of changing direction, meeting customer needs for new products and services, and staying ahead of the competition. Unfortunately, the perception and image of a public sector organization is that when a government executive approves an innovation project, it is usually a reaction to an external event or crisis rather than a result of insightful planning.

But, as events have shown in recent years, leaders in the public sector are also concerned with how to invest in smart city innovations, anticipate risks and threats, identify citizen-centric opportunities, and implement improvements and digital capabilities in their organizations and operating environments. Agencies at all levels of government are confronted with a more complex range of challenges and in some cases have fewer resources to meet the challenges. Therefore, the public administrator—like his or her private sector counterpart—has a continuing need to develop and maintain innovation management strengths.

Drivers of innovation in the public sector include:

- Changes in expectations of citizens and other stakeholders
- Pressures to improve crisis-management capabilities (in particular the ability to respond to economic and financial crises)
- Requirements to prepare for emerging threats (such as health, environmental or security crises)
- Opportunities resulting from smart city technology trends
- Staffing trends and shortfalls in government agencies
- Poor quality and breakdowns of antiquated government systems and procedures
- Pressures to serve citizens more effectively (through reform and improvements in public safety, intelligent mobility, education, economic development, health care, tax and legal systems)

Government Agencies: Catalyst or Barrier to Innovation?

Government entities should not be overlooked as potential partners and stakeholders in innovation projects. Depending on the nature of the project, government entities may serve in the role of funding source, sponsor, provider of research and knowledge, or occasional facilitator (as in the case of economic development agencies).

Long before the Internet and electronic mail reached a level of acceptance in the business world, the US Government was developing and using electronic message networks (such as the store-and-forward AUTODIN*) for non-voice messages and was responsible for funding of the ARPA** network, an Internet precursor for government and academic use. These were examples of government entities that were ahead of the business world in terms of telecommunications and network innovations!

In attempting to build knowledge societies, many countries have taken steps to modernize their national information and technology infrastructures. Their challenges—in acting as innovation enablers—include finding a balance among the dynamics of:

---

* Automatic Digital Network
** Advanced Research Projects Agency

- Implementing appropriate regulatory measures, while removing controls that inhibit innovation
- Ensuring universal access to networks and technologies
- Building indigenous technology skills and capabilities
- Promoting a competitive climate
- Deciding on where and how to apply public funding and research
- Protecting intellectual property rights

More recently, the bureaucratic hurdles presented in many government agencies have been exposed through problems that affect society and remain unsolved. Even in these cases, the government body that acts as a safety watchdog or regulator is also a stakeholder with interests in innovation and change. The commercial innovator (especially those with interests in digital government, technologies for public safety and national security, and smart city services and solutions) should therefore look toward the government stakeholder as a source of funding, research, project sponsorship, and regulatory and policy change.

A country's innovative strengths depend to a great extent on how participants in the economy (or within specific economic clusters) interact as part of a system of knowledge creation and application. The participants include established companies, startup enterprises, government agencies, government-funded research centers, universities and trade associations. The links among participants include joint research ventures, licensing agreements, mobility of technical personnel, and actions of trade associations to influence standards bodies and education initiatives.

Innovation at National and Multinational Levels

Successful countries have innovation characteristics in common, including openness to international cooperation and above-average innovation performance due to:

- Development of national innovation strategies and systems
- High investment levels in education, R&D and information technologies
- Government funding for R&D and innovation initiatives
- High share of business financing for R&D
- A diversified base of innovators and entrepreneurs

- Extensive networking and knowledge-sharing among innovators; with vibrant links between universities and industry

- Cooperative, innovative economic clusters

Through the strategy of national innovation systems (NIS), governments worldwide realize that an economy's innovative capacity is the result of complex interactions and relationships among the groups that produce and apply various types of knowledge.

NIS Success Stories

Although there is no standard definition of an NIS, it is generally viewed as a systematic web of interaction concerning knowledge development, knowledge sharing and diffusion, and technology transfer. The acceptance of the NIS concept is based largely on the fact that several countries—such as South Korea, Singapore and Taiwan—succeeded in transforming their countries from slow growth, trailing-edge laggards to positions of *high growth, high tech economies.* The NIS strategies of South Korea and Taiwan resulted in extraordinary programs of *intensive learning* (a critical success factor).

In addition to building internal research and technology skills, South Korea was highly innovative in terms of unbundling the technologies it acquired from external sources. Unbundling a technology involves separating each of the components and exploring how to supply the elements locally, perhaps in an improved form. "Components of disembodied technology can be disaggregated into basic process patents, basic designs, detailed engineering and specific engineering services. Technical assistance for the startup of operations, for example, and certain services such as detailed engineering or specific engineering could be available with the help of domestic design, engineering and consultancy firms." (UN, 1990)

South Korea's learning process enabled the country to develop capabilities for innovation in imported technology, processes, equipment, and standards.

South Korea–NIS Success Story

Despite dire initial conditions several decades ago, South Korea accomplished rapid and sustained economic growth after the 1960s, resulting in GDP per capita increasing more than eleven-fold. This was a unique occurrence on the world stage in the 20th century.

The accumulation of knowledge was a major contributor to Korea's long-term economic growth. The successful transition to a knowledge economy required long-term investments in education, development of innovation capacity, modernization of the information infrastructure, and creation of an economic environment advantageous to market transactions.

As an example of South Korea's commitment to development and innovation, consider the investment in information technologies. During the 1990s, the country invested heavily in its information infrastructure. A distinguishing feature of this investment was that a high percentage came from government. In many OECD countries, the public share in ICT investment was zero in the late 1990s. In South Korea, 25 percent of total investment was from the public sector.

Source: Worldbank, 2006

Enduring Features of US Innovation Strategy and Leadership

While various studies refer to a gradual decline in innovation strengths of the US in recent years, the fact remains that the US, over a period of several decades, continues as a global leader in many unique aspects of innovation:

- High percentage of immigrants founding and managing successful technology startup firms

- High level of government research and commercialization

- Abundant supply of venture capital for entrepreneurial ventures

- Mobility of workers and knowledge across industry, company and geographic boundaries

- Unusual culture of innovation and attitude toward risk-taking (e.g., treating entrepreneurial failure as a valuable learning experience)

- Propensity and freedom for innovation to flourish throughout all levels of government and industry (i.e., without constraints from a rigid bureaucratic authority)

Breakthrough! Innovation Management in Practice

Starting from the late-1940s, Defense-related R&D funding served as a catalyst for development, technology transfer and commercialization of civilian technologies in the US. The vital role of new firms in this process was unique among major industrial economies and was enabled by a domestic financial system which encouraged and facilitated formation of new high-tech enterprises. (Mowery, 1998) This process and structure for research and innovation continued through the decades.

Another unique feature of the US innovation system was that no central authority was formed to create or implement a national strategy. Instead, many Federal government departments, each with their own mission, culture and plans, forged their own research initiatives (Owen, 2016). Through the decades, this produced a diverse mix of R&D programs and led to a significant increase in the numbers of highly skilled and knowledgeable scientists, engineers, technicians and entrepreneurs in multiple industries. It also led to the growth of California's Silicon Valley and an unprecedented surge in the number of well-funded high-tech startup firms eager to take advantage of innovations and inventions emanating from government and academic research.

As discussed in Owen (op. cit.), "diversity and competition are hallmarks of the US innovation system—among funding agencies, among universities that compete against each other for talent and for funds, among innovation clusters such as those based in San Francisco and Boston and among firms."
In recent years, the US recognized its recent gradual decline and identified the actions needed to re-build and restore its strengths as a leading innovative nation. As described by the US Department of Commerce (2012), "the scientific and technological building blocks critical to our economic leadership have been eroding at a time when many other nations are actively laying strong foundations in these same areas." The DOC report recommended US policy decisions and actions such as:

- Continuing to support government funding for basic research
- Enhancing and extending the R&D tax credit
- Facilitating the transfer of innovation from science labs to commercial applications
- Increasing the spectrum for wireless communication
- Coordinating federal support for smart manufacturing
- Strengthening efforts to foster regional clusters and entrepreneurship
- Promoting efforts to improve access to foreign markets

A source of uncertainty for the future of innovation involves trends toward restricting immigration. US firms, especially those in technology-centric sectors, have relied on an influx of immigrants with high-tech skills, and this is often mentioned as a key success factor for many startups.

According to Anderson (2016): "Immigrants play a key role in creating new, fast-growing companies, as evidenced by the prevalence of foreign born founders and key personnel in the nation's leading privately-held companies. Immigrants have started more than half (44 of 87) of America's startup companies valued at $1 billion dollars or more and are key members of management or product development teams in over 70 percent (62 of 87) of these companies. . . .The collective value of the 44 immigrant-founded companies is $168 billion."

Innovation in the EU—Catching Up with the US

To increase innovation strengths in the EU, the European Commission (EC) released a report in 2002: "Innovation in a Knowledge-Driven Economy" and established the following priorities for promoting higher levels of innovation in EU member countries:

- Ensure coherence of innovation programs and policies
- Provide regulatory frameworks that encourage innovation
- Advance the creation and growth of innovative enterprises
- Improve key interfaces in the innovation system
- Promote a society that is open to innovation

Within a few years, the EC developed methods to evaluate innovation strengths and weaknesses of European countries and began publishing an annual *European Innovation Scoreboard* (EIS). When the EC published its initial EIS reports, the results were not favorable and indicated the EU was lagging behind the US and Japan in most key innovation areas.

During the next several years, Germany, Denmark, Finland, Sweden, Switzerland and the UK emerged as innovation leaders in Europe (based on EIS measurement methods). Switzerland in particular was successful in stimulating its innovation intensity through far-reaching government actions which demonstrated the commitment of this small country:

Breakthrough! Innovation Management in Practice

- Funding for national research programs–to provide for fundamental research in advanced materials and develop public and private research centers of excellence

- Development of networking programs–including national competence networks for science and materials research

- Small business incubation policy–enabling early-stage companies to operate inside established organizations (to design new products, establish contacts, and explore market opportunities)

The actions taken by Swiss leaders built upon the country's traditions in precision instruments and watch making and contributed to the country's economic and competitive strengths. By 2009–after evaluating innovation performance of all European countries on the basis of key indicators covering seven dimensions of innovation—the EC rated Switzerland as one of Europe's top five innovation leaders.

The EIS 2017 report stated the EU is catching up with the US, while the performance gap with Japan and South Korea is increasing. Denmark, Germany, Finland, the Netherlands, Sweden and the UK were rated as the EU's innovation leaders, and Switzerland (not a member of the EU) was Europe's most innovative country.

Although the EU has increased its overall innovation performance in the past several years and expects to reach parity with the US within a few years (according to the EIS performance measurement system), it remains to be seen when (and if) Europe can produce the innovation capacity, culture of innovation, venture capital funding and risk-taking mindset of the US. These are the ingredients that may ultimately determine Europe's ability to create, fund and nurture entrepreneurial, high-growth companies such as the global innovation leaders (e.g., Apple, Google, Facebook, Microsoft, Amazon, Tesla and others) established in the US.

The Czech Republic—Example of an Emerging Innovator

The Czech Republic (rated as a moderate innovator in the EIS 2017 report, with an innovation performance at 90 percent of the EU average)—with a population slightly above 10 million and a relatively youthful economy after its re-emergence in the 1990s—serves as a good example of a small country with ambitious plans for innovation. The Czech Republic's goal is to become a leading research nation and has a wide-ranging strategy to accomplish this vision, including investments in R&D, innovation and technology transfer, combined with participation in European Framework Programs and other research.

Adding to its growing strength in life sciences, electronics, software, precision engineering and sustainability, the Czech Republic is building upon a history of innovation and knowledge in manufacturing, vehicle design, glass production, and other important sectors.

One of the challenges for the Czech Republic is how to capture the value of an expanding body of knowledge by communicating an image of leadership in the international business community. The Czech Republic's research and innovation initiatives include:

- Investments in Extreme Light Infrastructure (ELI), a pan-European research initiative to design and produce breakthroughs in laser technology

- Implementation of a Center for Biotechnology and Biomedicine (BIOCEV)

- Development of a leading research center in Prague, to specialize in applied science in the fields of chemistry and physics

- Development of five research centers of excellence

- Application of resources from the European Research Area (ERA)

In addition to building these knowledge centers and innovation capabilities, another value generation strategy—although perhaps not as evident—is to apply thought leadership principles to take advantage of the investment in knowledge and research. A comprehensive thought leadership program designed to convey the unique strengths and knowledge in the Czech Republic would contribute to the Country's objective of becoming a leading research nation (Glasco, 2009).

Looking beyond the R&D results, a program of national thought leadership gives to potential investors and partners a better understanding of the nature of opportunities emanating from the research. As stated by George S. Day, op. cit. (Wharton), "when managers consider a new technology, they need to focus on the ultimate potential of the technology, and not be blinded by its current look and feel or the current shape of the market. The market will change. Technology will evolve."

For most smaller countries and regional economies in Europe, the advantages deriving from large population, large consumer markets and an export surplus do not exist. But small country and regional innovation programs do exist, and some of these are being aggressively implemented.

In Wales for example, government and business leaders believe small and medium-sized enterprises (SMEs) in the electronics, optoelectronics, tourism, automotive components, software and other clusters hold significant potential for growth and regional competitive strength. In contrast to Switzerland, which had the long tradition of strengths and skills in precision technologies which could be adapted to similar *strategically important* industries, Wales had to transform itself from an economy dependent on smokestack industries to something new. Wales had to identify which of the strategically important opportunities to pursue.

One of the major hurdles in this endeavor was that the Welsh employment base did not offer skills or knowledge for the business development and innovation that entrepreneurs and small enterprises required. It is one step to identify the opportunities, but another equally difficult challenge to find the skills and knowledge to pursue them. The task for the Welsh Assembly Government and other UK agencies was to develop and fund cluster-driven strategies that enable small businesses to grow, elevate employee skills, build innovation management strengths, and create sustainable positions in target markets. Actions taken in Wales included[*]:

- Forging alliances with foreign companies to win high added-value inward investment and transfer global best practices into Wales

- Promoting the exploitation of new technologies developed by Welsh businesses and universities

- Investing in high-tech incubators

- Establishing a fund to accomplish major changes in entrepreneurship and commercialization of new knowledge

The experience in Wales demonstrates that developing strengths as a regional innovator is often a long-term process. Because Wales was relatively poor (i.e., GDP per capita less than 75 percent of the EU average), the region qualified for more than GBP 3 billion of EU funding (allocated over a decade)

---

[*] Source: "Wales for Innovation: Action Plan for Innovation". (circa 2003). Welsh Assembly Government, Economic Development Department

to stimulate the economy and invest in research and innovation in SMEs. Despite this funding allocation, "studies have shown there is no real evidence of a step change in innovation capacity and performance in Wales, mainly due to the failure to address the disconnectivity between public sector funding and private sector interests." (Gkikas et. al. (2013), citing Jones-Evans and Bristow (2010) and Morgan (2012)).

According to Jones (2016), Wales still "lags behind the UK average on most measures of economic performance. . . . However, a focus on human capital (particularly addressing low skills), connective infrastructure (internal and external), and diffusing knowledge and innovation among firms should form a central part of the policy mix, and benefit the economy of Wales over the longer term."

Innovation Management at Local Government Levels

To confront emerging pubic sector challenges in the 21st century—and to address old challenges in new ways—local governments are searching for innovative solutions. In terms of service quality, urban renewal, e-government, sustainability and other challenges, citizens and businesses have high expectations that changes will be accomplished in government processes, public services, technologies and infrastructure—and these changes will deliver quality-of-life benefits for all segments of society.

"Managing urban regions has become one of the most important development challenges of the 21st century. If well managed, cities offer an incredible opportunity to generate economic development and to expand access to basic public services such as education and health care for large numbers of people, becoming healthier and more productive." (Torres, 2017)

However, most government leaders and citizens and business leaders agree there's a long way to go from today's challenges and plans to desirable visions of the future. Local governments face pressures to deliver more services, a higher level of quality, and a wider range of services. In addition, they must develop innovative and equitable solutions, improve efficiencies, and ensure a high quality of life in their communities. Urban planners must consider—in light of trends toward "smart city" development—how to plan for digital transformation of cities, towns and regions. (Toppeta, 2010)

Until recently new technologies and innovative concepts now receiving significant media attention, such as autonomous vehicles, robotics, 5G communications, Internet of things (IoT), predictive analytics, artificial intelligence, smart infrastructure, blockchain, and smart city platforms, were not on the urban planner's radar screen. "By leveraging digital and network technologies, cities can tap the collective intelligence of its many citizens and stakeholders, to work for solutions to urban problems, co-create new activities, and engage with them in many different ways." (Torres, 2017)

Preparing for and managing these complex changes are fundamental parts of managing local and city governments. How communities finance and accomplish future changes and how they transform and renew themselves depends on strategic decisions made by government and academic leaders, civic-minded citizens, local businesses and other stakeholders.

In considering the planning and decision-making functions of local governments and urban planners, a few questions arise such as:

- How do city governments and leaders use their websites to communicate with citizens, businesses and other stakeholders about:
  - Plans for an integrated smart city (e.g., encompassing smart, integrated systems for transport, energy, security, education, governance, environment, utilities, water and sanitation)
  - Smart city civic engagement program or activities
  - Availability of a smart city strategy (or similar publication)
  - Quality of life commitment
  - Smart city collaboration or partnerships

- When city governments perform strategic planning (or other transformation planning), what is the focus of their plans?

- Are the strategic and long-term planning challenges in city governments similar to those in the private sector?

To address these questions, I conducted a Smart Trust Survey in early 2017, to identify plans and strategies of 45 city governments. I searched for and examined information regarding smart city visions, urban strategies, and plans for transformation. Although not a scientific survey, my goal was to take an online snapshot of a wide range of local governments to explore the nature of their plans and how they communicate with citizen stakeholders.

This survey yielded the following results:

Mentioned / included in city website?	Nr of cities
Integrated smart city strategy or plan	0
Smart city civic engagement	9
Smart transport vision (response to US Department of Transportation 2016 Challenge)	12
Technology initiatives with smart city capabilities	40
Availability of smart city strategy publication	0
Quality of life commitment	33
Availability of apps with smart city features	19
City executive or manager with smart city responsibilities	5
Smart city collaboration or partnerships	20
Involvement with smart city vendors	10

Based on the survey results, it appears few cities use their websites to communicate or clarify smart city strategies, plans and civic engagement. Technology initiatives were mentioned (by 88 percent), along with availability of apps with smart city features (by 42 percent) and smart city collaboration (by 44 percent), although none of these were clarified in the context of an integrated smart city strategy or plan.

Seventy-three percent of cities mentioned their commitment to high quality of life. However, none of the city websites defined—in an explicit manner—smart city strategies to improve (or maintain) quality of life, except for the smart city transport visions and various data initiatives, which were not described as part of a smart city plan. None of the city websites mentioned whether they have a smart city strategy available (e.g., in the form of a strategy publication) to citizens and other stakeholders.

None of the city websites in the survey mentioned smart city (or other technology-centric) projects involving a breakthrough or major improvement in innovative public services.

A complex issue for government leaders is that quality of life means different things to different people. And it may have different definitions from one city or state to another. According to the International Economic Development Council (IEDC), "quality of life is the economic well being, life style, and environment that an area offers. Improving the quality of life is the ultimate aim of economic development programs and initiatives. A balance has to be maintained between encouraging the growth of the local economy, while limiting impacts upon the quality of life. In this post-industrial new economy, people are increasingly seeking better quality of life."

Quality-of-Life Dimensions

Employment opportunities

Quality and accessibility of transportation

Quality of education and lifelong learning

Medical facilities and health care services

Affordable housing

Environmental protection

Low pollution levels

Public safety and security

Recreation, sports and entertainment

Aesthetic build and natural environment

Perhaps in the era following the economic crisis of 2008 and 2009, many citizens expected their public sector leaders to have a more participatory and innovative role in preventing future economic meltdowns.

Most city governments provide information on their website about capital budgets, capital investment plans, and financial management. This is important information to deliver to citizens and businesses, because it demonstrates that government is making an effort to provide the community with a desired level of services as a means of maintaining or improving quality of life.

The multi-year capital investment plan in most cases is used to confirm the need and funding levels for replacement of aging infrastructure to maintain traditional service levels or in some cases to improve infrastructure prior to service modernization.

Most local government capital budgets present the funding requirements for services in the upper left-hand and upper right-hand quadrants of Figure 9-1 (adapted from Glasco, 2012).' In terms of urban and regional planning, this is the *low-hanging fruit.*

Figure 9-1. Innovation Management in Local Government Planning

		Public Services Demand	
		Established / Old	**Emerging / New**
Local Gov Delivery Options	**Traditional (low risk)**	Maintain services Maintain quality Maintain efficiencies	Service modifications Improved quality Increased productivity
	New / off-the-shelf (moderate risk)	Incremental changes Proven applications Infrastructure changes	Service innovation E-gov services New applications
	Transformation (higher risk)	Not applicable	Smart city / digital services Equitable smart city solutions Infrastructure modernization Digital infrastructure Intelligent mobility Re-invention of urban space Digital inclusion

However, many of the problems resulting from the 2008 and 2009 economic crises, combined with geopolitical and other challenges in the years following, point toward the need to consider new planning quadrants in the "emerging / new" column of Figure 9-1. In the lower right-hand quadrant, radically new (breakthrough) services, infrastructure and systems to meet new needs are required to improve (or ensure) quality of life. Few city government plans explain how new services are considered, planned or funded.

Building Value Through Government Transformation

In many respects, leadership and management challenges at the city government level are similar to those encountered by business leaders. Most city government leaders, including mayors, city council presidents, council members, city managers, departmental managers and urban planners face the same types of challenges encountered by CEOs and other corporate executives.

They must plan and manage traditional functions and activities such as:

- Financing
- Human resources
- Safety and security
- Service creation, delivery and quality
- Information and telecommunication systems
- Infrastructure investments
- Cost of operations
- Tradeoffs between short-term improvements versus long-term investments
- Stakeholder engagement

In comparison with government, private sector leaders are usually more concerned with the need to transform their organizations by responding to and anticipating changes in markets, customer demands, competitive actions and legal and regulatory trends. For a professional planner in private enterprise, the terms strategic planning, transformation and innovation management are all closely linked. The planner serves decision makers who are concerned with value generation activities, i.e., creating and delivering value to customers and positive returns to shareholders.

Government leaders and urban planners also strive to create value in their communities—in addition to their stated priorities of addressing the basic service mandate and other goals (restoring economic health and ensuring sustainable development for example). Especially for the long term, a government's transformation initiatives are a prerequisite to creating value and improving and maintaining quality of life.

In a public service situation, value is a highly complex mix of tangible and intangible components. A local resident or business owner can describe what they expect from government in terms of traditional services: law enforcement, public education, public libraries, land management, zoning regulations, parks and recreation, fire department, water system, and public transit services. All these services and more, when delivered to meet basic needs, contribute to the citizen's quality of life. When the citizen believes quality of life is high, then the citizen values the community and develops a sense of pride in living and working there.

Value creation and the quality of life it delivers should never be taken for granted. Over a period of time, the value (perceived or actual) of a community may go up or down. As in the corporate situation, changes in value result from how decision makers develop and implement strategy and how they plan and invest in the future.

In some cases—as Peter Drucker stated (op. cit.)—"the strategy itself is the innovation." Some services, such as those delivered by a municipality, have been around a long time. The city government strategy could serve to transform an established service into something new. As Drucker pointed out, the strategy changes the utility of the service, its value, and its economic characteristics. Adapting Drucker's theory to local and regional challenges, the urban planner has opportunities to develop and apply strategy for:

- Creating or increasing utility
- Developing innovative financing and revenue generation models
- Adapting to the citizen's social and economic reality
- Delivering real value to the citizen

The government leader therefore needs a way to *measure value generation*, even if some parts of the equation are difficult to quantify. Unlike the corporate situation where value often refers to a financial return, the generation of value in a community refers to quality-of-life benefits plus a sense of optimism for the future and pride that results from shared principles, beliefs and standards of conduct.

Following is a proposed value-of-community (VOC) metric:

$$VOC \approx \frac{QOL + QLS + OPF + TLL}{CLG}$$

Where:
QOL = quality of life
QLS = quality of local services
OPF = optimism for the future
TLL = trust in local leaders
CLG = cost of local government

Figure 9-2 provides a suggested model of value creation through government transformation.

Figure 9-2. Local Government Transformation Value Cycle

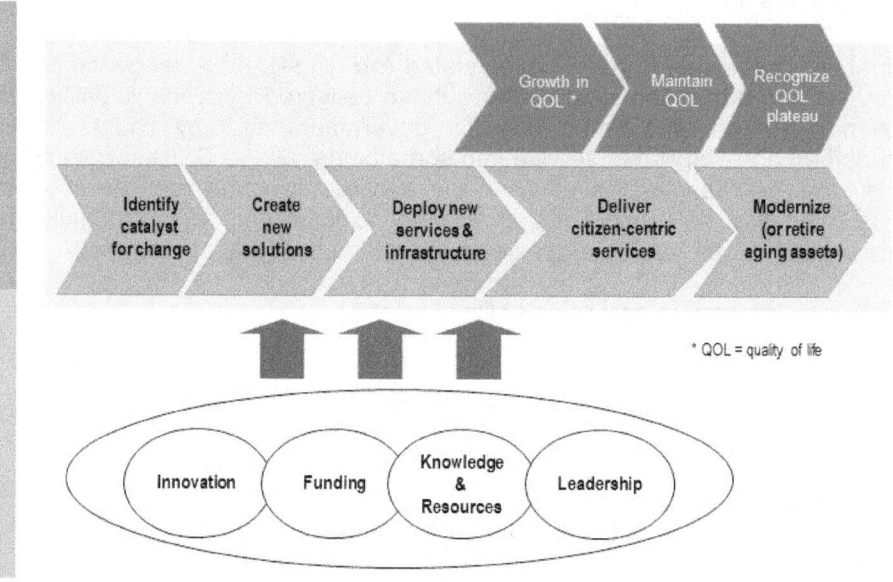

In this model, the value cycle begins with recognition of a catalyst for change, which might include any of the following issues and challenges:

Low quality of public services
Demand for smart city services
Local housing shortages
Unacceptable crime rates
Education shortfalls
Transport congestion
Unaffordable broadband
Public safety problems
Aging population

Climate change and environmental risk
Threat of natural disaster
Social exclusion
Demand for e-government
High unemployment & poverty
Low digital literacy
Need for resilience
Demand for digital governance
Need for improved emergency services

After recognizing catalysts for change, the value cycle continues with creating and designing new urban, rural and regional solutions; deployment of new services (and new infrastructure if required); delivery of citizen-centric services (designed to serve all segments of society); and eventually to modernization (or retirement) of services and supporting infrastructure. In parallel with the community's value cycle, the quality of life (QOL) should improve with development and launch of new services and public initiatives.

However, the measured and perceived QOL will eventually reach a plateau as societal needs change, and this should reveal signs of a pending QOL slump. This is a signal to urban leaders and planners that it is time to modernize (or retire) services and infrastructure and begin a new transformation cycle.

Influential Stakeholders in Innovation

Through their decisions and actions (or inaction), public policy makers and other institutional stakeholders have a considerable impact on innovation. Stakeholders who influence national, regional and urban innovation include:

- Government departments that influence public policies, R&D strategies and budgets, technology transfer, and support for a pro-innovation environment
- Trade commissions and world trade organizations
- Regulatory and standards development bodies
- Intellectual property entities
- Legislative bodies, courts, tax and competition authorities
- Public safety and environmental watchdogs
- Urban planning, regional planning and economic development agencies
- Universities
- Trade associations

In a study of businesses in Canada, Dalziel (2005) compared the impact of various external organizations on innovation capabilities. The study's premise was that a company's ability to innovate is enhanced by the exposure to new ideas from external sources.

Customers, suppliers and competitors were mentioned most often as significant sources of ideas. But among the non-profit entities—government, universities and trade associations—the latter was more influential as a source of ideas.

This is not surprising, given the increasing numbers of associations in many industries and the wide range of services they deliver to their members. For example, the services offered by trade associations in high-tech industries include:

- Organizing and presiding over seminars and lecture series
- Providing training and career development services
- Publishing white papers, policy recommendations, technical books, and other thought leadership materials
- Accelerating the adoption of technologies
- Facilitating a knowledge exchange for resolution of technology and product implementation issues
- Promoting the commercial value of technologies
- Assisting members with market development, trade and export promotion, standards issues, and domestic and international advocacy
- Developing implementation agreements, promoting worldwide interoperability, and encouraging input to standards bodies

Although the term *innovation management* may not be mentioned in their mission statements, the diversity of activities and support makes the trade association a partner, enabler and facilitator of innovation.

Alliance for Innovation—A thought leader and partner in local government transformation

The *Alliance for Innovation*, based in Phoenix, Arizona is an organization dedicated to innovation and transformation in local government. As described by its President, Karen Thoreson: "The Alliance connects innovative and successful local governments in robust communities of practice and shares new research and resources with our members who strive to excel in all endeavors."

Serving as a vital resource for knowledge and emerging practices in local government, the Alliance:

- Organizes and sponsors an annual *Transform Local Government* (TLG) conference to promote networking, team building, and opportunities to share knowledge and ideas

- Convenes an annual invitation-only event (*BIG Ideas*) that gathers progressive leaders to explore critical issues for the future of local communities

- Maintains an extensive *Knowledge Network*—an online community for local government professionals offering content and thought leadership in all facets of local government management and operations

- Participates with Arizona State University and ICMA (International City/County Management Association) in the *Enhanced Research Partnership* —which developed a free civic engagement tool for local governments to understand their readiness and capacity for engagement

- Operates an *Innovation Academy* with courses and workshops on understanding the process and culture of innovation, building cross functional teams, cultivating leadership, and increasing organizational capacity for innovation

Source: Alliance for Innovation | Transforming Local Government
http://transformgov.org/en/home (accessed on 10 July 2017)

Breakthrough! Innovation Management in Practice

Economic Development Agencies and Innovation in Clusters

Countries and cities throughout the world have recognized the importance of innovation in developing and maintaining economic strengths.

There is also an understanding that geographic proximity of many firms in the same industry enables a form of networking, communication and collaboration that benefits all firms in the sector. This is another example of how a stakeholder acts as an innovation enabler by serving as a *cluster facilitator*. Essential tasks for the cluster facilitator are to encourage, promote and facilitate networking activities through symposia and workshops to address industry-wide issues.

Another task for the cluster strategist / facilitator is to develop and promote an understanding of how the innovative style and strengths of the country or region affect innovation management objectives.

Economic Clusters and Shared Paradigms

One of the challenges faced by an economic cluster strategist is that the companies who participate in a cluster assume diverse roles: competitors, partners, suppliers, and/or customers. And each company has a unique mission, vision and strategy.

Therefore, the concept of cluster participants working together to form a cluster strategy may not have meaning or value to some of them. They may consider the cluster as potentially damaging or threatening, perhaps due to information that must be shared. They may not trust other members of the cluster, especially those with whom they have competitive rivalries. In this situation, the cluster facilitator must build trust among the members and create the *common ground* where all participants appreciate the value of cooperation. This common ground represents a shared paradigm of success.

Figure 9-3 depicts the types of enterprises in the public transport industry that might work together toward a shared paradigm in their cluster.

Figure 9-3. Shared Paradigm in a Public Transport Cluster

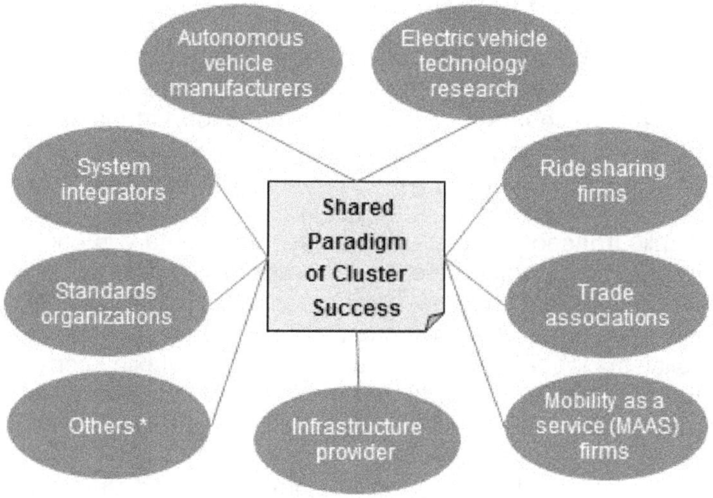

* Others: Public transit agency, urban planners, universities, data scientists, EHS (environment, health and safety) firms, user experience (UX) specialists

The cluster facilitator helps team members collaborate on a range of strategic options and avoid debating the details of implementation. The implementation comes later, and depending on the nature of innovation challenges is accomplished at the individual firm level or by business partners or through government-sponsored research.

Success factors for the cluster strategy and its implementation include:

- Willingness to collaborate and share information
- Commitment of regional or local government to provide funding and other resources for cooperative projects and partnerships
- Opportunities for collaboration and clarification of the benefits through a shared paradigm of success
- Communications and trust among cluster members
- Persuasive communications to influence government and promote policies that encourage entrepreneurship and innovation
- Champions and owners for knowledge-sharing projects

In the life cycle of most economies, there are occasionally periods when companies, governments, universities and innovators have opportunities to accomplish major changes. Sometimes these opportunities are driven by societal changes or needs to solve a difficult problem pertaining to the environment, national security, crime, education or transport.

The US public transport industry in the early 21st century is one of the sectors with wide-ranging opportunities to innovate and accomplish breakthrough changes for the benefit of cities, rural areas, transport operators, passengers and society in general. However, most transport projects depend on favorable public policies, leadership and funding. This public commitment from government and other innovation enablers (such as universities) in turn depends on awareness of the socioeconomic benefits of such projects.

Therefore, to pursue the opportunity, companies and entrepreneurs in the transport sector need to build *thought leadership* in regard to the socioeconomic benefits. These benefits—such as improved safety and security measures, protection of the environment, reduction in urban congestion and noise levels, and improvement in quality of life for all segments of society—are difficult to quantify. And in most countries and urban areas, new concepts in the transport sector are judged attractive only after detailed analysis and preparation of a business case that confirms the profitability and *business value* of the venture.

Adding to the complexity of proposed changes, most breakthrough solutions for public transport problems have additional challenges (beyond the business case) in terms of quantifying socioeconomic benefits.

Companies and startup ventures seldom have the resources to provide research and analysis on the potential benefits and costs of decisions that affect socioeconomic trends. This is an example of how the plans and funding for breakthrough solutions—and in some cases the health of an industry— often require actions among diverse stakeholders at multinational, national or regional levels:

- Development of close ties among universities, companies and government that allows research organizations to clarify the benefits and costs of proposed projects

- Risk sharing among funding sources, suppliers and transport operators (with recognition that new concepts will entail more risks, and suppliers should not bear the entire risk burden)

- Development and facilitation of clusters of companies, government entities, thought leaders and academia

- Streamlined regulatory measures which recognize that some aspects of breakthrough solutions and technologies (such as autonomous vehicles and robotics) must by necessity challenge existing standards, while demanding that changes in the regulatory sphere occur almost in parallel with design and test activities

- Metrics which allow measurement of socioeconomic aspects of projects during project implementation and after new technologies, products and services are operational

Communications with Stakeholders: Building Trust in the Strategy

> *"The larger the organization, the more difficult it can be to deduce precisely what it does simply by listening to what it says."*

> Mick Herron, The Language of Business

A responsible and dedicated city government works to build trust in the community. This is accomplished to a great extent through the messages and methods of communicating with citizens, local businesses and other stakeholders.

Therefore, it is surprising the words *innovation* and *transformation* seldom appear in local government websites. None of the city websites in the *Smart Trust Survey* offered an explicit transformation goal or statement nor did they mention or provide access to a municipal strategy publication.

As discussed in Goldsmith and Crawford (2014), Mader (2016), MRSC (2016) and Glasco (2017), local government leaders have opportunities to build civic trust in strategies and innovations that affect the long-term future of communities by communicating and engaging with citizens through:

- Websites, website extensions and social media

- Public hearings, town hall meetings and virtual meetings

- Community workshops and forums

- Co-creation and participative democracy

- Problem-solving workshops

Breakthrough! Innovation Management in Practice

- User experience (UX) testing
- Digital governance methods and tools

"Building and maintaining trust in smart city plans, projects and solutions should be part of the urban innovation process and involve all partners in the process including government, citizen groups and task forces, urban planning consultants, smart city vendors, universities, non-profits and other stakeholders." (Glasco, 2017)

Perspective on Crafting the Future

From the 1960s through the mid-1990s, many changes in government and industry were almost linear. Technology changes occurred at a rapid pace, but most other aspects of managing a large enterprise and responding to external trends were relatively predictable. Starting in the 1990s, the widespread use of Internet-related technologies began to create non-linear disruptions in the world. Since September 2001, the world has learned that geopolitical tensions and terrorist attacks produce other unpredictable shock waves. As a result, the process of transformation demands new skills in creating and managing a climate for change, innovation and stakeholder communications.

Transforming city government requires looking beyond the replacement of aging infrastructure or improvements in existing basic services. It demands creation of new services for new situations. It demands new thinking and new methods for crafting the future of communities and adding value to the quality of life through equitable smart city solutions. A transformation value model serves as a planning tool and facilitates the public dialogue for city governments.

In retrospect, leaders in public and private sectors should also consider how to apply modern innovation management practices to the spectrum of issues and questions that societies, governments and people are concerned about—especially in view of economic and financial meltdowns of 2008 and 2009. It seems obvious that we need to change the outdated systems for deciding, financing and controlling how governments, financial institutions, regulatory bodies and citizens plan for the future—to ensure the power for transforming communities and crafting the future resides in the hands of the many—rather than in the narrow corridors of power controlled by the few.

10 | Fusion of Innovation and Strategy

"Most innovations fail. And companies that do not innovate die."
Henry Chesbrough

Innovation does not happen by chance, or if it does, the nugget of opportunity may be overlooked due to other priorities and pressures. As described by PwC in their *Innovation Benchmark Study*, "random acts of innovation rarely pay off. For any initiative to deliver true value, the effort must clearly align with a company's business strategy. Yet, successful alignment between innovation strategy and business strategy can elude even the best of companies." (Staack and Cole, 2017)

This chapter discusses how to craft the fusion of innovation management with the process of strategy development and implementation.

Innovation, Strategy and Risk

The process of developing and implementing a strategy—although defined in analytical terms and complex quantitative models—is also a creative activity.

Accomplishing a strategic transformation involves a high degree of visionary skill, change, uncertainty and risk taking. Supplier relationships, market and product strategies, partnerships, processes, internal systems, digital platforms, and perhaps the business model itself are modified or replaced.

In many companies, the process of creating strategy is an exciting time. It is a time when grand visions are formed, new opportunities discovered, promising ventures considered, and innovative—perhaps revolutionary—products are envisioned. Presentations and workshops are designed to be informative and motivational. Managers and employees are encouraged to get creative and think laterally! Although competitive threats and barriers to success are considered in the strategic planning process, the general tone of workshops and strategy documents is hopeful. Optimism prevails.

So it is a dilemma that the risks inherent in managing a business lurk beneath the programs needed to implement the strategy. The task of confronting risks acts like an anchor on the momentum required for successful implementation. Therefore, this task is usually delegated, perhaps under the guise of risk management or another *behind-the-scenes* role.

The risks posed by the marketplace, regulatory authorities, financial markets, customers, fraudulent activities, political unrest, and a wide range of other forces are real and—more often than not—cause damage to strategies and tactics. Too often in past years organizations reported poor results and explained that "unforeseen events" or other external forces (such as competitors' aggressive moves or the shock of disruptive technologies) had an adverse impact on performance. History is littered with the remains of companies that simply did not recognize or understand the risks inherent in their strategic decisions and business environment.

Risks in Abundance

Numerous companies and government entities lament that—yes, the world is a more uncertain and risky place than ever before. "Risk management failures at major corporations have captured the headlines for many years, primarily in the financial sector, but in other sectors as well, and have not always been the result of shortcomings in financial risk-taking. Environmental catastrophes such as Deep Water Horizon or Fukushima come to mind (or, less recently, Bhopal and Seveso), as well as accounting fraud (e.g. Olympus, Enron, WorldCom, Satyam, Parmalat), or foreign bribery (e.g. Siemens) cases, to name just a few from the non-financial sector." (OECD, 2014)

But the days when companies could refer to events beyond their control as the cause of poor performance may be over. Executives cannot rely on risk management resources to provide all the answers or avoid all the risks. However—with or without adequate risk management tools—organizations must place assets at risk to achieve objectives.

The Risk of Failure

Corporate history demonstrates that strategic initiatives taken by organizations often carry a high risk of failure, whether starting a new venture, making an acquisition, developing new technologies and products, or entering new markets. Recognizing this as an inherent part of corporate life and taking the steps to innovate in spite of risks is a characteristic of companies that achieve breakthroughs.

Apple Computer serves as an example. In a period spanning decades, Apple has been admired for their development and introduction of popular technologies and products such as the first Mac computer, Mac OS software, MacBooks, the iMac, the iPod and iPhone.

However, in this same period the company suffered innovation problems with many products that have mostly been forgotten: the Lisa, Pippin and Apple III. Also forgotten or overlooked is that the value of innovation is difficult to capture if the implementation does not meet customer needs. An example was the ill-fated Apple Newton, an early form of personal digital assistant, which was viewed by many as a legendary product blunder. However, according to Dodgson and Gann (2010), some of the technologies and features designed for the Newton were re-applied (and the value captured) in the highly successful iPod and iPhone product lines.

For Apple, the benefits of breakthrough successes far outweighed the costs of failed products. In spite of product development mistakes, management changes, loss of engineering talent, and severe competition, Apple's climate for innovation and design excellence survived, providing an example of how a healthy climate (established as an integral part of the culture) is sustainable over the long term.

While executives sometimes blame external circumstances for strategic and business setbacks, I have observed six strategy shortfalls which—acting alone or in combination—can lead to failure.

Six Strategy Shortfalls to Avoid

Lack of strategic planning practices or processes for implementation

Poor understanding of information, ideas and internal knowledge about potential opportunities

Inability or lack of commitment to integrate strategy with a modern innovation management process

Lack of commitment from key managers and employees to take the pioneering steps required to challenge competitors and status quo conditions

Attempts to implement innovative measures without the new structure and resources needed for success

Inflexible plans (which hinder mid-course corrections when unexpected situations arise)

Table 10-1 summarizes risks which should (with help from innovation management) be addressed in the strategic planning process.

Table 10-1. Major Categories of Strategic Risks

Risk Category	Risk / Uncertainty	Potential Impact
International concerns	Unstable conditions: macroeconomic, geopolitical or other global upheaval	Currency volatility Political instability International tension & conflicts Terrorism / cybersecurity threats
Process issues	Mistakes in deploying resources to implement strategy or accomplish transformation	Lack of commitment to strategy Loss of market share Process cost & complexity Errors or delays Partnership disputes
Competition & reputation	Events or mistakes concerning markets, competitive strengths & customer perception	Erosion of customer confidence Decrease in sales & profitability Disruption of third-party relationships (e.g., digital platform suppliers) Damage to fulfillment capabilities Data privacy shortfalls
Macro-environment threats	Losses due to vulnerability caused by external factors	Disruptive market changes Financial market crises Disruption of operations (due to industrial or nuclear accidents, earthquakes or other disasters) Unfavorable regulatory actions
Tangible asset complexities	Mistakes associated with acquiring & managing assets	System, safety or security failures Investing in the wrong technology Capacity excess or shortfalls Failure to recognize tech disruption
Intangible asset issues	Misjudgments in managing human resources, intellectual assets & soft skills	Dysfunctional workplace Damage to brand & reputation Loss of leadership talent Lack of key skills in workforce High employee turnover

The era has passed when management of strategic risks could be delegated to an accounting department or risk manager or simply included in the company's insurance policies. When an organization finds itself confronting unexpected threats, management must be prepared with appropriate contingency measures, resources and role clarity.

Dealing with Risks: A Process Within a Process

Based on experience with Fortune 500 companies, it is apparent that executives, board members, parent organizations and shareholder companies do not need a strategy *carved in stone*—i.e. an inflexible plan that establishes a course of direction without a mechanism to respond to changing conditions. What they do need is a strategy (with flexible contingencies) capable of adapting to new opportunities, threats, and the risk of unforeseen circumstances.

When problems and hurdles emerge, innovation management expertise adds value by *creating a process within the strategic planning process* to identify risks, options, problem-solving measures and strategic changes. To develop this expertise, management must ensure innovation management practices and associated resources are part of the strategy development, documentation and implementation process.

One of the reasons for developing a strategy is to show how the organization will create value within a future environment where it will operate. Whether in the present or future, the environment is inherently risky. Risks that are not understood—leading to unanticipated changes—may damage the organization's ability to create value and capture that value in the form of profit or other result measures.

In successful organizations, the strategic planning process relies on both analysis and insight (i.e., left-brain and right-brain thinking about the future) and produces a flexible plan and culture—with the ability to adjust to evolving situations. Ensuring that the innovation management process, associated resources and tools are integrated with strategy development and execution should provide these strengths.*

---

* For a wide-ranging list of strategic planning and decision-making tools, refer to University of Cambridge, Information Technology Management:
http://www.ifm.eng.cam.ac.uk/research/dstools/

Breakthrough! Innovation Management in Practice

**Integrate Innovation
Management
and
Strategic Planning**

Develop strategic risk assessments, identify major categories of risks, and generate ideas on how to deal with the consequences of risks

Clarify how situational changes (in markets, customer demand, technologies) could adversely affect the execution of strategy and generate ideas on how to anticipate and monitor changes

Use scenario planning to assist in building a dialogue on the risk of disruptive forces

Develop methods for monitoring the business environment to increase sensitivity to subtle changes

Create *agile response* plans—to provide rapid and flexible methods for proactively responding to threats, opportunities or other changes

An agile response plan, also known as a living document or *adaptive strategy*, is capable of adjusting to changes in the organization's environment and enabling decision makers to alter structure, internal systems, partnerships, supplier arrangements, prices and other strategy levers.

Adding Value through Strategic Communications

It is inevitable for most companies that development of strategy often requires planning a new direction for the organization, finding new ways to set the company apart from its competitors, and identifying what to do in the future that was not done in the past.

Another reality is that while executives and planners are developing the new strategy, other managers and employees throughout the organization are still committed to the old strategy and continue implementing that strategy. Executives must inform and persuade them that a new strategy has emerged. How, when, where and how often to communicate will become critical to success of the new strategy. One of the main reasons why companies encounter problems in strategy implementation is the lack of employee, engagement, understanding and commitment.

Therefore, strategy documents and implementation plans must include persuasive communications—through the company intranet, internal newsletter, informal meetings, electronic mail and training. Persuasive communications adds value to the strategic planning process by:

- Clarifying the strategy and its implications for change
- Engaging with employees to build trust in the strategy
- Explaining how the new strategy should be implemented
- Revealing the benefits of a favorable outcome
- Encouraging quality of communications among stakeholders
- Adding to the company's base of intellectual capital

Creating Strategic Alternatives

In many organizations, the starting point for strategic planning activities is a series of strategy workshops. These are usually high-level conferences, perhaps convened at off-site locations where ideas are presented and discussed concerning:

- How to redefine or improve the business model
- How and where to pursue new opportunities and enter new markets
- How to address competitive threats and other risks
- How to improve the yield of R&D or other major investments
- Where and how to make changes in business operations
- Whom to consider for strategic alliances and partnerships

As ideas are generated and analyses performed, a range of new strategic alternatives will sometimes emerge. The crafting of these alternatives demands that innovative energies be brought to bear on the opportunities, threats and challenges. Therefore, it is worthwhile to explore how the process of innovation management should be linked with strategic processes. Unfortunately, new opportunities and emerging threats—which may imply changes in business strategy and direction—are often neglected or rejected because of perceived barriers or risks.

According to Chesbrough (2006), an organization which rejects an opportunity later implemented successfully by others is committing a Type II error, a "false negative". On the other hand, a Type I error, the "false positive", is when the organization pursues an alleged opportunity (usually within its established core business), that fails to meet expectations.

Chesbrough describes an organization with a tendency to commit Type II errors as "not a good poker player" and therefore not as effective in addressing situations with lack of information and many unknowns. Therefore, attractive opportunities are seldom discovered or considered in these organizations. In such circumstances, strategic planners need *exploratory thinking* methods to discover and present innovative alternatives.

The Strategic Options Matrix in Figure 10-1 (adapted from Royce, 1979) is a method for crafting *strategic alternatives.* For each decision category in Figure 10 (numbered 1 through 6, which might include decisions on markets, products, pricing, technology, partnerships and regulatory policy), a list of options are created.

Figure 10-1. Strategic Options Matrix

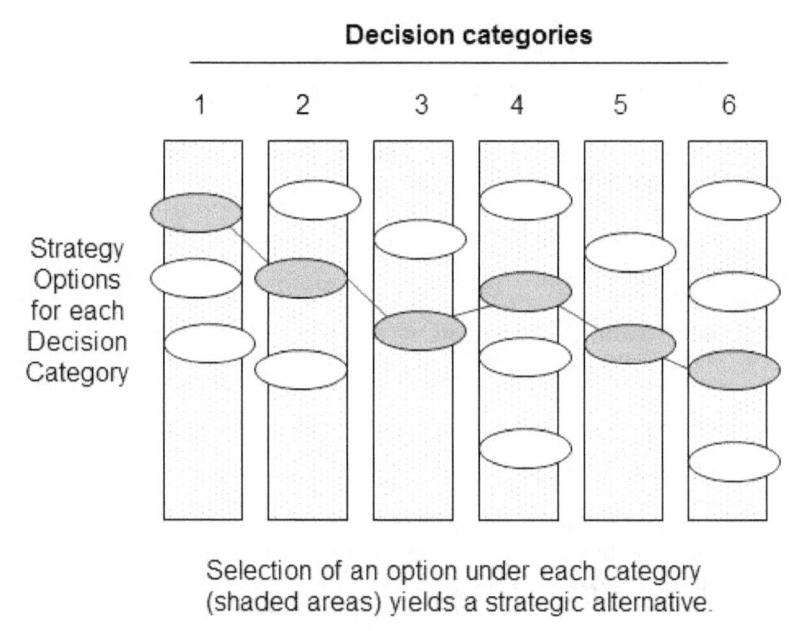

Selection of an option under each category
(shaded areas) yields a strategic alternative.

Innovation workshops provide the climate and resources to further develop each option, identify potential benefits, and discuss measures to mitigate the inherent risks and avoid Type II errors.

The advantage of using the strategic options matrix as a method for crafting strategic alternatives is that it promotes discussion and innovation regarding the decision categories. Team members are encouraged to place their ideas and views under each decision category.

By using this method, the facilitator (a) demonstrates the volume of ideas (an indicator of progress); and (b) guides the process toward the structured creation of a manageable set of alternatives.

After the options are developed under each decision category, it's time to look at the big picture—to craft a strategic alternative based on selection of one option under each category. In the figure, the choice of shaded options would constitute a strategic alternative. The alternative may need to be further crafted, refined and developed before subjecting it to a full-blown analysis.

As in all innovation projects, a facilitator needs to ensure that each strategic alternative identified has:

- A champion or owner
- Subject matter experts with an understanding of feasibility and implementation options for the alternative
- Sufficient definition and clarity to allow for analysis and refinement

Cooperative and Innovative Partnerships

Mergers. Strategic alliances. Joint ventures. These and other forms of formal and informal partnerships have been with us for a long time and for better or worse are part of the landscape of business, government and academia. But from an innovator's viewpoint, is a partnership a useful mechanism for achieving innovation? Do partnerships sometimes erect barriers that inhibit the innovative spirit?

Consider four partnership situations:

- Joint innovation ventures
- Joint ventures with multinational and multicultural dimensions
- Project-driven ventures
- Innovation-centric supplier agreements

Joint Innovation Ventures

In a strategic joint venture, two or more companies agree to share knowledge, technology, capital, processes, risks and rewards in the formation of a new entity.

The lure of a strategic *joint innovation venture* offers compelling benefits, such as the potential for:

- Combining complementary R&D activities
- Merging unique technologies to develop a breakthrough
- Acquiring marketing or distribution expertise
- Commercializing a technology, product or business concept
- Achieving critical mass in target markets
- Expanding into new markets
- Accelerating revenue growth and increasing profit margins

A success story from the mobile phone industry involved Sony Ericsson, a 50:50 joint venture of Sony Corporation and LM Ericsson, that was established in 2001. Up to this time, LM Ericsson was at the forefront of cellular phone innovation and reached a decision to obtain mobile phone components from a single source, a Philips-owned facility in New Mexico. In early 2000, a fire at this facility contaminated the sterile conditions needed for production and compromised product delivery schedules for months.

Faced with serious shortages, Ericsson's development and launch of new handsets was adversely affected. After serious losses, the company decided to change its mobile phone business strategy, and the partnership with Sony was formed.

Sony Ericsson introduced its first jointly developed products in 2002 and established a full and competitive product portfolio. In 2005, Sony Ericsson introduced the Walkman-branded music phones—recognized as the first music-centric mobile phones. The venture's innovative momentum continued with the introduction of the CyberShot™ and Experia™ brands and product lines.

This venture demonstrated that design and development can be accomplished through multinational cooperation taking place in multiple locations. The Sony Ericsson Creative Design Center established studios in Japan, Sweden, the UK and US with industrial designers and other specialists involved in a wide range of innovative design activities.

.

Key Factors to Consider in Creating a Joint Innovation Venture (JIV)

Screening and due diligence of prospective partners

Knowledge, assets and other resources to contribute to the partnership

Communications with staff and employees to build support for the JIV and create a climate for collaboration and innovation

Development of a business plan which includes:

- Definition of the JIV objectives
- Expectations for a breakthrough or other innovation
- List of prospective partners
- Legal structure of the partnership
- Provisions for intellectual property protection
- Methods for sharing responsibilities, costs, risks and rewards
- Exit strategy

Joint Ventures and Multicultural Challenges

In spite of admirable intentions, partnerships are sometimes not highly conducive to innovation. Some of the toughest partnerships to manage, in terms of achieving strategic objectives, are those involving multinational and multicultural situations. In spite of the difficulties, these types of joint ventures are popular, because they offer partners the possibility to exploit their strengths in a mutually beneficial manner. A foreign firm can minimize the risk of investing in a new, overseas market, and the local partner firm benefits by gaining access to new technologies and global markets.

In spite of the growing popularity in this form of cross-border partnership, many such ventures have proven to be unstable and not capable of meeting objectives. Some of the steps in building a successful multicultural joint venture include:

- Developing the JV organization plans and publishing plans in the language of each partner
- Clarifying roles and responsibilities
- Ensuring that project leaders are aware of partner expectations toward innovation, risk and knowledge sharing

- Clarifying how the cultural values of each partner's employees are valued and protected when planning and implementing the venture

- Creating a platform for communications that allows partners to share information about issues and problems

- Providing facilitators from each partner organization (conversant in the partner's language)

Project-Driven Ventures

Companies in diverse sectors such as civil engineering, consulting, software development, film production and others depend on project-driven ventures as a source of revenue and value generation. Project-driven ventures offer the benefits of bringing unique knowledge and diverse talents into a sphere of interest where opportunities are known to exist. The project team—if given the resources and process and unencumbered by the rule-based structure of other organizational forms—has the potential to deliver high value in a relatively brief period of time. However, project-driven ventures present unique innovation management challenges due to their chaotic nature. Team members usually may not have a well-defined scope; roles and responsibilities are in a state of flux during various project phases; participants often do not have long-term working relationships or trust; and sponsors may be wary of the project's feasibility, costs and delivery estimates.

An example of project-driven complexity comes from the transport sector, where development of passenger rail vehicles is a market dominated by a handful of major manufacturers. Each of the major manufacturers is both a project-driven enterprise and system integrator. To secure business, the system integrator first submits a bid document (for the design and production of vehicles) to a rail operator, public transit authority or other government entity.

Each bid, and ultimately the engineering work performed by the winning bidder, depends on a complex matrix of equipment suppliers, subsystem contractors, and others that provide the vehicle components such as car shell, electrical system, air comfort system, brakes, couplers, vehicle monitoring system, and numerous others. It is the responsibility of the winning bidder (the system integrator) to bring together all the systems from multiple suppliers and assemble, test and deliver the family of new vehicles.

The passenger rail business dates back to the 1800s. We might expect that designing vehicles is a straightforward process, proven by decades of past success.

However, in many rail vehicle procurements, the customer defines distinct (and often unique) requirements regarding vehicle performance, safety, passenger comfort, quality, maintainability, training, documentation, and other attributes.

The winning bidder faces a system integration challenge that gets more complicated if the customer specifies innovation requirements which strain existing practices or standards—or if the customer specifies a radically new type of subsystem or performance level. In such cases the system integrator will search through its supplier base to find someone capable of meeting the customer's unique requirement. Several suppliers may be asked to submit proposed approaches to the design and production.

The system integrator will then conduct design meetings with one or more preferred suppliers, and after selecting a supplier for the work will start the process of engineering the specified subsystem. This is where problems often occur. Following are examples of the issues that may arise:

- Lack of trust in sharing information on sensitive technologies and processes
- Poor technical communications among project team members
- Differences in interpreting the specification
- Lack of experience in cooperating across organizational boundaries
- Cultural differences in terms of confronting and managing risks
- Incompatible technologies for interfaces between the new subsystem and others with which connects
- Incompatible project management systems

Making the Supplier a Partner in Innovation

With the recent trend toward outsourcing, companies are faced with the challenge of finding ways to ensure suppliers are partners in innovation. This requires suppliers, especially those in the small and medium size (SME) category, to alter their established practices for purchasing agreements, which are often more adversarial than cooperative. And the buyer must be prepared to share not only the details of its requirements, but also open its technology planning and product development process to supply chain partners.

**Get Suppliers Involved
Early in the Process**

A key issue is deciding when to bring suppliers into the innovation process. In most cases, the earlier the better.

This helps in making supplier personnel feel they are part of the innovation team and motivates them to address the challenge in creative ways.

As the team and process move closer to implementation, suppliers will have more in-depth understanding of what's needed to ensure success in meeting (or exceeding) the buyer's expectations.

Suppliers must in turn be willing to come forward with ideas, concepts, and products, while delivering new insight and innovation that reach beyond the buyer's minimum requirement.

A model of successful *system architect and supplier partnerships* comes from Cisco Systems. Cisco demonstrated the power of partnerships when they proved they could compete with the world's major telecommunication vendors. Whereas many of the leading vendors had for many years invested in closely protected internal R&D, Cisco's strategy was to create a more open innovation process, do less internal research, and instead rely on external partners. Cisco succeeded in forming an innovative global network of partnerships to build its technology and product strengths.

Although these strategic moves by Cisco, combined with proprietary patents and unique software, enabled the company to establish long-term market leadership, competitors responded by releasing innovative products with open-source code (which allowed independent software firms and partners to develop user applications at lower costs). This demonstrates that innovating and competing through networks of supplier-partners is a continuous process.

Table 10-2 provides examples of actions and objectives that foster open innovation and promote cooperative partnering relationships with suppliers.

Table 10-2. Buyer-Supplier Innovation

Stage of innovation process	Actions	Objectives
Issue identification	Supplier days E-collaboration	Develop conversation on customer challenges and problems Build trust
Ideation	Supplier days Off-site workshops E-collaboration	Share ideas on possible solutions, technologies and product development Build trust
Concept development	Invitations to presentations Joint design activities Direct communications (including intranet)	Prototypes and technical issues understood by all Improve understanding of next steps and responsibilities Knowledge sharing
Test and Deployment	Joint development activities Shared platforms Long term contracts Marketing cooperation	Share risk and cost management Meet performance and delivery standards

Merging Strategy Development with Innovation Management

Although many companies and government organizations emphasize the necessity and inevitability of change, there are signs that making the requisite changes are often difficult.

Unfortunately, change management and the innovation management process are not always integrated with the process of strategy development. The innovation management process should function alongside and in support of business planning and strategy development. Table 10-3 provides examples of how innovation management adds value in the strategy development process.

123

Table 10-3. Value of Innovation Management (IM) in Strategic Planning

Strategic Options	Implications & Potential Benefits	Hurdles & Risks	IM Deliverables
Joint ventures and mergers	Risk sharing Access to resources and knowledge	Difficulties in achieving synergies	Build trust Shared paradigm of success New knowledge, new opportunities
Acquisition	Access to new markets and knowledge Achieve rapid growth	Difficulties in integration Hurdles to capturing value	Facilitate continuity among planning, deal making and integration Build trust Transform tacit knowledge into explicit
Market development	Leadership potential First mover advantages	Market uncertainties Competitor moves	Ideas and concepts from IM process Identify new opportunities
Product development	Leadership potential	Demand uncertainties Technology hurdles	Ideas and concepts from the IM portfolio Facilitate new product idea generation and knowledge management
Strategic partnerships	Technology transfer R&D sharing Cost sharing	Complexity of implementation Trust building	Quality of communications Facilitate knowledge sharing interfaces
Diversification	Move away from traditional core businesses	Market uncertainties Places other units at risk	Ideas and alternatives concerning options (horizontal versus conglomerate; how to exploit channels)

Breakthrough! Innovation Management in Practice

Concluding Remarks on Innovation Management and Strategy

As Peter Drucker advised (op. cit): "purposeful, systematic innovation begins with analysis of the opportunities." When conducting an *opportunity assessment*, innovation management practices ensure that opportunities for incremental improvement and potential breakthroughs are part of the strategic planning portfolio.

When an organization faces a new or threatening situation and solutions are limited, *innovation management provides tools* to facilitate the development of a wide range of strategic alternatives. This enables people to look beyond the horizon of a crisis and turn problems into opportunities.

To deliver capabilities for responding to crises and changes in the external environment, innovation management practices should be applied to the process of scenario planning and development of an *agile response plan*.

To contend with strategic and operational risks, innovation management offers an *exploratory approach* to planning. Combined with facilitative and analytical tools, this enables innovation management professionals to (a) imagine and visualize how situational risks or disruptive forces might adversely affect the organization's future, and (b) clarify the options for avoiding or mitigating the risk factors.

Persuasive communications techniques heighten strategic planning efforts by clarifying the strategy and its implications while showing how an innovative strategic path leads to a favorable outcome.

Knowledge management—including *knowledge sharing and development of new ideas from existing knowledge*—are success factors in the process of innovation. A body of research has emerged that highlights the importance of networks in knowledge sharing activities, in both inter- and intra-organizational forms of networks. According to Fagerberg et al. (2005), the knowledge-transfer process has at least two distinct approaches. One approach for the flow of information and ideas through networks involves the reliance on "complementary assets in the division of innovative labor." A second approach occurs when "existing information within a network is recombined in novel ways."

If information in networks is tacit and offers high value, an innovation manager can facilitate the flow of such information and its ultimate conversion to more explicit, transferable knowledge, thereby increasing the knowledge sharing capabilities of the network.

When integrated with strategic planning *innovation management adds value* by delivering new ideas and alternatives, building trust and communications among stakeholders, and facilitating knowledge-sharing interfaces. Ultimately, strategy and innovation, working in harmony, should serve as a guide to discovery in uncharted territory.

Historians tell us that Christopher Columbus, mariner and explorer, ushered in the Age of Discovery. As the historian John N. Wilford wrote in *The Mysterious History of Columbus*, "exploration has been called planned discovery." In this sense, the concept of managing an innovative enterprise has been around for centuries—since the late 1400s when Prince Henry the Navigator demonstrated his skill as an innovator by developing the concept of nationally-sponsored, high-risk exploration by sea. At about the same time, the Italian geographer Toscanelli was gradually replacing medieval geography with a different view of the world that suggested the "shortest sea route to Asia lay to the west, straight across the Ocean Sea." Columbus assimilated this knowledge (i.e., how to secure a sponsor for his exploration) and new vision of the earth and its oceans to plan his voyages to the New World. (Wilford, 1992)

At their respective moments in history, Prince Henry and Columbus were acutely aware of the risks of transoceanic exploration. Perhaps, in our moment in history, the modern practice and techniques of innovation management will enable us—when we confront risks and unknown horizons—to discover new worlds of opportunities and solutions.

Appendix A–Innovator's Self Assessment

How to Give Your Organization an Innovation Management Health Check

The astute leader who is concerned about building and maintaining innovation strengths will benefit from an occasional *health check* of innovation management strengths and weaknesses.

An organization with a healthy approach to innovation and change is one that implements innovation management processes and capabilities:

- Fully supported by senior management
- Driven by a strategy of innovation leadership
- Led by a champion of visionary services
- Staffed with professionals trained in innovation management techniques
- Integrated with other planning processes (such as strategy development, market plans and project plans)
- Supported by non-traditional role capabilities
- Supported and trusted by employees and other stakeholders

A recommended self-assessment of innovation health is provided in the following questions. If your average score on this assessment is 3.0 or above, then most likely you have a healthy and stimulating climate for innovation.

	4–Strongly agree
Metrics	3–Somewhat agree
	2–Somewhat disagree
	1–Strongly disagree

My organization and its leadership:	*1*	*2*	*3*	*4*
Ensures we have an internal climate and culture conducive to change and innovation				
Searches for new opportunities on a continuing basis				
Encourages employees to learn from all potential sources, including informal, social and professional networks				
Encourages trust between and among groups and individuals				
Promotes and rewards *exploratory thinking*				
Is committed to quality of interpersonal and interdepartmental communications				
Is committed to *knowledge investments*				
Recognizes that employees must sometimes take risks in presenting new ideas				
Ensures strong linkage between strategy development and innovation processes				

My organization and its leadership:	1	2	3	4
Relies on innovation management principles and best practices to accomplish change				
Encourages and engages employees to get involved in innovation projects				
Recognizes that employees must sometimes assume *non-traditional roles* on innovation projects				
Rewards employees who take part in innovation projects				
Recognizes barriers to innovation and is committed to removing the barriers				
Encourages and nurtures breakthrough ideas and concepts from all levels of the organization				
Recognizes the potential value of open innovation (including ideas from suppliers, customers and other external stakeholders)				
Recognizes that not all innovation projects succeed				
Recognizes that innovation often depends on a new process or changes in existing processes				

Appendix B–The Language of Innovation

Absorptive capacity
Ability to recognize and absorb new knowledge and emerging ideas and apply these to the challenge of innovation.

Additive comparison
A tool used for comparing alternative concepts and solutions; helpful when the ratings of multiple criteria are in different dimensions or scales. Relies on converting the ratings to dimensionless values (which are then multiplied by a weighting factor to reach a final score for each alternative).

ADI: Asset-driven innovation
A process of finding new ways to create value from existing or to-be-acquired assets. Involves the use of untapped asset value as an *entry point* for ideas.

Participants in innovation are encouraged to explore new opportunities for both *tangible assets* (R&D infrastructure, production facilities, working capital) and *intangible* (unique skills, management practices, favorable contracts, partnerships, path dependency advantages, patents, copyrights, customer databases, etc.) (See: *Entry point*)

Alternate choice technique
A persuasive communications technique which involves suggesting two equally acceptable choices to the listener; helps to avoid a "no" option during early-stage idea generation activities.

Analytical hierarchy process (AHP)
A model created by Dr. Thomas Saaty as a strategic decision making aid; used for comparing decision elements which are difficult to quantify. [Refer to Saaty, 1980]

Ansoff matrix

Market and product innovation tools (created by Igor Ansoff) used to help planners consider alternative growth options via existing and/or new products and markets. Four possible product / market combinations help planners decide what strategy and actions to take:

- **Market penetration**–existing markets with existing products
- **Product development**–existing markets with new products
- **Market development**–new markets with existing products
- **Diversification**–new markets and new products

Aspiration level

A specific outcome that an individual or group wants–usually based on a perception and mental model of risks, costs and benefits. For a given situation, people will have different models and therefore different aspiration levels. In an innovation project, it is important for the facilitator to have an understanding of conflicting aspiration levels. (See: *Mental model*)

Assumption reversal

An idea generation technique where key assumptions regarding a particular scenario are listed, and then reversed before addressing the same scenario to discover what new solutions emerge.

Attribute listing

A problem-solving technique that consists of listing the main attributes or characteristics of a scenario or problem and then analyzing each attribute with the goal of trying to change them in multiple ways.

Barriers to innovation

Obstacles and problems that hinder attempts to innovate. Includes technology hurdles, cost overruns, lack of knowledge, funding shortfalls, poor climate for innovation, risk-averse culture, regulatory barriers, lack of commitment.

Brainstorming

Method for generating a large volume of ideas and concepts quickly and spontaneously in a mostly unstructured meeting; sometimes called an ideation session. Success of the method depends on having a facilitator, a good climate, and the ability to suspend judgment on the ideas produced.

Breakthrough

A concept or solution that is "new to the world" and reaches beyond anything that's been accomplished before. May involve high risk and often requires exceptional persuasive communication skill to secure approval for development. (See: *New-to-the-world solution*)

> "Discovery consists of seeing what everybody has seen and thinking what nobody has thought." Albert Szent-Györgyi de Nagyrápolt, Hungarian physiologist, Nobel prize winner, 1937

Camelot scenario

A technique to help innovation teams think beyond the status quo. Participants are asked to create an ideal solution or situation (the "Camelot") and compare this with their current situation or problem, e.g., what are the differences? Why do these exist? How should we design an ideation session to create new opportunities and solutions that bring us closer to the ideal scenario? [Higgins, 1994]

Capitalist

A capitalist is skilled in locating and securing funding and other resources needed by innovators; the commitment of the capitalist may be an assumed, non-traditional role (beyond the traditional job description).

Climate for innovation

The internal environment and attitude toward innovation and risk-taking; critical to the success of innovation projects. Participants in such projects must believe they are encouraged to generate new ideas, think beyond existing boundaries, experiment with breakthrough concepts and at times take risks they would not consider in the normal course of their work.

Codified knowledge

Knowledge or information content that has been recorded or documented in some form of communications medium (printed document, engineering drawings, computer file, software, website content, blog, e-mail message, social and professional networks, intranet, video and audio recordings, multimedia presentations, digitized images).

Consensus building

A group-oriented approach to planning and decision-making that seeks agreement of all or most participants and the step-by-step resolution of issues and objections.

The approach aims to be inclusive, participatory and solution-driven. However, it is usually not a good approach when an innovative solution is required and may hinder development of a healthy *climate for innovation.*

Concept development

An early phase of the innovation process that involves transforming ideas into realizable (and fundable) concepts; often relies on experimenting with concepts through the use of models and prototypes with the objective of making the ideas more concrete and acceptable to decision makers.

A process facilitator works toward building a shared paradigm of success in this phase. (See: *Shared paradigm*)

Connection mode

A creative process of visualizing something new by connecting ideas and thoughts which were previously considered unrelated.

> "It is the tension between creativity and skepticism that has produced the stunning and unexpected findings of science." Carl Sagan, American astronomer, astrochemist and author, 1934-1995

Creative extraction

Mixing together two or more different ideas or models to extract, as if by smelting, a totally new concept or model.

Decision tree

A planning tool based on a tree diagram that serves as a model of alternative decisions and helps to estimate and clarify consequences (risks, costs, profitability, utility, etc.). Used to identify the alternatives most likely to achieve a desired objective.

Generally not used in the early stages of an innovation project, but can be helpful after concepts are selected for implementation and need to be compared prior to committing resources.

Delphi method

A highly systematic method of analysis and forecasting that relies on a group of selected experts who respond to questionnaires in multiple rounds. Following each round, an independent facilitator delivers an anonymous summary of results and asks experts to revise their earlier replies (based on replies from other group members). The process is designed to converge on a best solution or answer.

Applied as an innovation technique when the facilitator wants each expert to think through each round without immediate reaction from others.

Design thinking

A solution-driven way of creative thinking; starts with a goal or vision of the innovative outcome (rather than starting with a specific problem or opportunity). Explores the parameters of possible ways to reach the solution. Relies on synthesis (rather than analysis) of ideas and concepts.

Direct analogy
A problem-solving approach that looks for similar situations to the current one and asks questions such as: "Where and how has this problem—or something analogous—been solved in the past? What does the prior solution offer in terms of new ideas, entry points and knowledge?" (See: *Entry point*)

Diverse resource
Someone with a creative spirit; an articulate idea generator recruited from the periphery of the project or organization. Usually adds new perspective and fresh ideas from external sources.

Dominant position
An opinion or point of view by a person of authority; one who often controls the process of innovation (or may hinder the process through an unwillingness to change views or accept the ideas of others)

Ecosystem
A group of interdependent entities and their environment. Example of an *innovation ecosystem*: A pro-innovation network or cluster of suppliers, customers, business partnerships, R&D ventures, entrepreneurial firms, investors, universities, government agencies, and trade associations.

Embedded knowledge
Knowledge that becomes rooted in a company's products, services, systems or processes. A specialist applies expertise to create something of value that depends on what has been learned through experience and training.

Virtually all business and manufacturing processes are developed from what was once the knowledge (perhaps tacit) of individual experts.

Entry point
The selection of an issue or situation from which to generate ideas. Selection of an entry point is important in ideation sessions, because the sequence in which ideas flow may affect the conclusion and ultimate solution, even when the cluster of ideas is the basically the same. (See: *Task headline*)

Exploratory thinking
A process of thought that avoids evaluation and decision-making in the early stages. The goal is to experiment with ideas, sometimes fail, learn from the experiment, and keep going toward other, untested ideas and concepts. (The opposite of a decision-mode style of thinking, which is appropriate when we need to take action or reach a decision quickly.)

> "If I find 10,000 ways something won't work, I haven't failed. I am not discouraged, because every wrong attempt discarded is another step forward." Thomas A. Edison, US inventor, 1847-1931

Facilitator

Someone who serves as a guide for the innovation process; plans and facilitates team meetings, ideation sessions and problem-solving activities. May also be in charge of process design. Usually is not involved in content.

This role is indispensable in most innovation projects. The facilitator builds and maintains a healthy climate for innovation and ensures new ideas are encouraged and nurtured.

> "If you put forward a new idea, someone will immediately tell you what is wrong with it and why it will not work—a sort of management by exception, totally misapplied. We pay an enormous price for this cultural quirk." Nolan, p.62

Fishbone diagramming

A technique to (a) identify all the possible causes of a problem or situation; and (b) conduct ideation sessions to address the main causes. Helps innovators to examine the total problem space rather than looking at a narrow cross section.

Groan zone

A state of mind between divergent and convergent thinking; where a group experiences frustration and discomfort with the process of innovation and / or their role in the process. [Kaner, 2007]

Headlining

Method of using a brief phrase or sentence to present an idea before giving any supporting details; helps to focus attention on the nature of the idea without getting distracted.

When the climate is less than ideal, many idea generators will have a tendency to pre-sell their idea before revealing it. This causes the listeners to focus on details, and they may not hear the idea when it is presented.

The quality and quantity of ideas increases when participants apply the headlining technique with ideas often expressed as "I wish" or "How to" statements. (See: Synectics®)

Hidden position

A potentially innovative and valuable position or viewpoint held by a team member or functional department that lingers in the background of innovation projects; often kept in the background and not revealed to the project team due to perceived risks, fear of failure, or other barriers.

Idea champion

Someone with an entrepreneurial mindset who promotes and encourages idea generation, finds ways to nurture and develop new concepts, and advises other innovators on both the content and process for innovation management success.

Idea portfolio

A structured knowledge base of ideas, concepts, unexplored opportunities, potential solutions and other intellectual assets that promote and stimulate exploratory thinking.

Idea profile

A brief description on the nature and characteristics of an idea, its positive features, potential value, areas of concern, and proposed next steps. The idea profile serves as a tool in clarifying the idea and building momentum.

Ideation

The idea generation and brainstorming phase of an innovation project; designed to obtain a wide range and large volume of fresh ideas to address complex challenges and problems.

This is a highly creative endeavor (rather than analytical) and depends on having a good climate for innovation and an understanding of *suspended judgment.*

A process facilitator works toward the generation and nurturing of ideas and avoids evaluation of feasibility or risk. (See: *Innovation project*

In / out listening

The mind wanders! It's not a bad thing, but most of us, when listening to someone else, will tune out occasionally. This happens because of the volume and diversity of information, pressures and priorities we have on our minds in a typical business day.

Synectics® teaches that in/out listening is beneficial when we recognize that it's happening. When we "check back in" with the speaker, it is often because a new idea has emerged in our mind, perhaps one that connects with something the speaker said earlier. The key is to write a quick note or reminder phrase on the idea before it is forgotten or obscured by other information.

Innovation

The process of creative thinking and development of new ideas and concepts, combined with opportunity assessment and implementation.

Innovation audit

A structured process of reviewing a completed project or process to identify what worked, what problems occurred, and generate ideas and recommendations for the future.

Innovation management

The process, tools, climate and resources needed to generate, nurture and develop new ideas, concepts and solutions—and to derive value from the results.

Innovation management process

The interconnected steps taken to achieve innovation; often designed and re-designed while a project is in motion. Success of the process often depends on non-traditional roles and the ability to maintain a favorable climate for innovation.

Innovation project

An assembly of resources, activities, knowledge sharing and ideas created to design and build something new. Most innovation projects are less confined than traditional, structured projects, and are usually characterized by high energy levels.

Intellectual asset management (IAM)

The process of creating, identifying, leveraging, managing and applying intellectual assets in ways that maximize value; involves management of both tacit and codified knowledge.

IAM is related to innovation management through the new opportunities that an enterprise pursues to capture the value of its intellectual assets. (See: *ADI*)

Intent versus effect

Sometimes what we intend to communicate is not what the listener hears.

In an innovation project, many new ideas and terms emerge, making it difficult for team members to communicate. When someone offers a new idea, the listener(s) may not understand the intent.

This is why it's important to use the *paraphrasing* technique as a means to communicate to the idea generator that we are making an attempt to understand the idea and the intent. (See: *Paraphrasing*)

> "The mismatch between intent and effect is a common cause of communication failure." Nolan, p.19

Intrapreneur
A creative employee in an established organization who has many of the characteristics of an entrepreneur. Takes responsibility for innovating and turning ideas into new products and solutions.

Inverse brainstorming
A form of idea generation that starts from the opposite side of traditional brainstorming. With the inverse method, a project team is given a new (perhaps breakthrough) solution and asked to look for potential problems in design, implementation, testing and marketplace acceptance.

Issue identification
One of the initial phases in an innovation project; when the goal is to identify problems, opportunities, threats, emerging situations or other drivers of innovation.

It isn't necessary to define issues in precise terms in this phase, but rather to compress the issue into the form of actionable "How to" or "I wish" statements—which encourage the generation of fresh ideas in the *Ideation* phase. (See: *Springboards and Task headlines*)

Itemized response
An idea-nurturing technique which calls upon a group to look first for the positive aspects of an idea and then convert its drawbacks into "how to" statements. The how-to's are used to invite further development of the idea and find ways to increase its acceptability. (See: *Synectics®*)

Kepner-Tregoe
A systematic process used to maximize the skills and knowledge of participants when addressing a specific problem, opportunity or decision. Also known as the 'Rational Process', it usually consists of several parts: Situation Appraisal, Problem Analysis, Decision Analysis, and Opportunity Analysis.

Knowledge investment
The use of money and other resources to develop or acquire knowledge; an ingredient of innovation management. May include investing in various combinations of:

- Technology transfer

- University collaboration

- Content management and digitization

- Knowledge management and knowledge-sharing initiatives

- Research and development

- Market and competitive research

- Thought leadership programs

- Mergers and acquisitions

Knowledge turn
The time required for an idea or experiment to proceed from initial hypothesis to results in the marketplace (Grove, 1998)

Lateral thinking
A form of thinking and reasoning about a problem or challenge in a manner that is not obvious or consistent with current perceptions; a method of thinking concerned with changing perceptions and mental models; generating ideas and solutions that are not within reach through traditional logic or analysis.

The term was originally created by Edward de Bono. (See: *Vertical thinking*)

> "In lateral thinking one does not mind being wrong on the way to a solution ... it may be necessary to go through a wrong idea in order to get to a position from which the correct path is visible." De Bono, p. 232

Leap-ahead thoughts
The mental phenomenon of listening and interpreting at a rate that exceeds the speaker's rate of communicating—often leading to a perceived conclusion that may not match the speaker's intent. (See: *Intent versus effect*)

Leapfrog
In idea generation, this is a logic-defying thought that enables innovators to mentally leap over assumptions or barriers which block the flow of and ideas.

Marketing innovation
New methods and improvements relating to the functions of pricing, marketing communications, promotion, distribution, packaging, and thought leadership.

Market stretch

Using innovation tools to develop new ideas for extending the life of a product or service.

Enables planners to generate proactive ideas to deal with an expected sales decline or loss in market share. Usually relies on entry points such as new packaging, new product uses, product enhancements, and changes in promotion and sales channels. (See: *Entry point*)

Mashup

Intuitive and creative blending of dissimilar concepts or ideas. Can be applied at various levels of innovation, e.g., product design, complex systems, business model, national and regional innovation strategies, etc.

Mental model

An individual viewpoint on how things work in an industry, market, organization or functional area. May be difficult to quantify or communicate to others. May hinder the process of innovation due to beliefs that *assumed barriers* cannot be overcome.

The role of a facilitator is to guide innovation team members by bringing mental models into focus so that barriers can be visualized, discussed in a constructive manner, and ideas generated for positive action.

Metaphor

A figurative use of language in which two different modes of thought or ideas are linked by some similarity. The metaphor treats one thing as though it were something else so that a resemblance or characteristic that would usually not be in view is highlighted. (See: *Direct Analogy*)

Through an *extended metaphor*, innovators are motivated to develop a new (primary) line of thought with multiple (subsidiary) ideas for further development.

> "All the world's a stage. And all the men and women merely players. They have their exits and their entrances." William Shakespeare | *As You Like It*

Mind map

A diagramming technique that displays words, ideas, tasks, or other items linked in a radial structure around a key word or idea. Image-centric diagrams are used to represent semantic or other connections between pieces of information. Used to generate, visualize, structure, and classify ideas, and help in organization, problem solving, and scenario planning.

141

Mini-dominate options
An approach to strategy when a company has limited resources and is unable to build a dominant market position.

The mini-dominate method encourages people to explore new ideas for concentrating their marketing efforts on selected market segments or geographic regions; sometimes combined with ideation sessions that focus on how to conserve and focus resources on the most attractive opportunities.

Minimalist outlook
A group or individual perspective that promotes simplicity in design. In an innovation project, the minimalist is someone with creative skills who is effective in seeing through the complexities of a situation and generating ideas for highly simple solutions, strategies, product designs, etc.

Minimax principle
A theory about decision-making that suggests individuals seek to maximize rewards and minimize costs.

In zero-sum game theory and statistics, the principle is thought of as minimizing the maximum possible loss. Alternatively, it can be applied to maximizing the minimum gain (maximin). Applications include complex gaming situations and decision-making under uncertainty.

Useful in the late stages of an innovation project to create a dialogue for discussion of uncertainties.

Morphology analysis
A problem-solving method that enables the generation of many ideas in a brief time span. The method relies on development of a detailed matrix with specific characteristics, adjectives and words related to the problem. The goal of the analysis is to force a set of characteristics against another to create fresh ideas and new insights.

Network architect
Someone who maintains awareness of key contacts and resources in adjacent networks of interest to an innovation project. Assists the project team in obtaining and absorbing knowledge and ideas from other networks.

Skilled at exploring new networks for relevant information and building information-sharing relationships and boundary-spanning linkages.

New-to-the-world solution
A completely new approach to a problem or opportunity; usually difficult to articulate or define in early stages; may depend on several innovations and inventions working in harmony. Typically a high-risk, high-cost and complex endeavor. (See: *Breakthrough*)

Non-traditional role
A function that someone performs which is outside their traditional job description or assigned role. On some types of innovation projects, a non-traditional role may not be officially sanctioned or defined, but is one recognized as essential to success of the project. May involve some degree of career development risk to the person assuming the role (but it is part of the project leader's role to ensure personal risks are minimized).

Offer
One of the five basic forms of innovation dialogue, as suggested by Synectics® theory. An 'offer' is an idea, suggestion or any other communication intended to add value.

The other four forms of innovation dialogue are:

- **Accept**–"I like your idea."
- **Reject**–"It doesn't fit with our strategy, and it's too risky."
- **Build**–"I like this idea, and here's another thought that adds to what you're suggesting."
- **Question**–Any probing communication (usually placing the idea generator on the defensive).

Open innovation
A concept and theory that promotes sharing of research, knowledge and innovation with external entities, including partners, competitors, customers and other stakeholders. [Chesbrough, 2003]

Opportunity assessment
The process of perceiving, detecting and recognizing opportunities for change, new technologies, products or solutions; may not depend on detailed analysis at first but instead relies on weak signals from the external environment. Often serves as the initial stimulus for an innovation project. (See: *Weak signals*)

Paradigm shift

In the context of innovation, the term is often used (and abused) in connection with a major change in thought and personal beliefs or a radical change in complex systems or businesses.

Paraphrasing

To paraphrase is to capture the meaning or at least the basic characteristics of someone's idea or viewpoint. To contribute to a healthy climate for innovation, it is often necessary to repeat the meaning of someone's comment or idea in our own words and not simply repeat their words verbatim. (See: *Offer*)

Parking lot

During an ideation session, this is an area where the facilitator records ideas and concerns that are tangential to the current session, but which have value for another session or phase of the project.

Persuasive communications

The process and practices used to develop messages, models and documents for a specific audience that the writer wishes to influence; relies on the knowledge and application of language, logic, evidence and information design. Documents that depend on persuasive communications include business plans, information memoranda, white papers, proposals and strategic plans.

Process innovation

Changes and improvements in an organization's process, for example in business operations, financial management, marketing, product development, business development, human resources management.

Usually concentrates on improving efficiencies and effectiveness, but may sometimes focus on creating a new form of business model or providing a mechanism for entering new markets.

Product innovation

Development of new products or improvements in old products and services. May involve creation of a unique mix of services to complement an old product.

Introduction of the new product or service may also provide opportunities for new approaches to market entry, partnerships, promotion, distribution channels, packaging, and thought leadership.

Problem-solution structure
A logical problem-solving approach and communications pattern that relies on a sequential process, which provides details on the problem, and proceeds through analysis and conclusions to ideas and recommendations. This structure is the opposite of lateral thinking and headlining techniques for presenting ideas. (See: *Lateral thinking*)

Project sponsor
Someone who provides the initial impetus for an innovation project and team formation; takes a major role in identifying the project goals, selecting team members, coaching the facilitators, and securing funding and other resources. May provide a vision of the desired outcome. (See: *Visionary*)

Protocept
An emerging product idea or concept that is crafted and refined through experiment and simulation (prior to prototype design).

Reversal method
A provocative method of creativity requiring the idea generator to take a situation as it is currently understood and turn it around; provides a means of escaping from the old way of looking at a problem or challenge.

> "By disrupting the traditional mode of thinking, one frees information that may come together in a new way . . . by making the reversal, one moves to a new position. Then one sees what happens." De Bono, p. 143

S-curve
The theory that most innovations spread through a market or society in the form of an S-curve, e.g., early adopters try the innovation first (market grows slowly along the curve), followed by a majority (rapid growth), and finally leading to acceptance by the mass market (growth plateau).

This theory was first introduced by E. M. Rogers who said a customer's attitude toward a new technology or product went through five stages: Knowledge, Persuasion, Decision, Implementation and Confirmation.

Scenario planning
A planning technique that relies on using available knowledge to create plausible future situations; sometimes used in conjunction with systems thinking methods. (See: *Market framing*)

Screening matrix
A method of choosing a group of ideas for further development; usually a two-dimensional matrix with 'attractiveness' along one axis and 'compatibility' (or another relevant measure) along the second axis.

Self censor
A pattern of negative thinking which causes an individual to critically evaluate emerging ideas which are internally generated; resulting in the ideas being suppressed (the opposite of *exploratory thinking*). The self censor is usually more prevalent when there is a poor climate for innovation.

Shared paradigm
A jointly defined model and dialogue where participants in an innovation project share a vision of success and achieve a high level of trust, knowledge sharing and cooperative energy.

Situational leadership
A theory of organization stating that different and flexible leadership styles and models are needed for different situations. In an innovation project, a leader with a quality-centric style ("get it right the first time") may have to adjust his or her style to confront the risks of innovation (when people rarely get it right on the first try).

Soft innovation
A form of innovation (usually not based on technologies) involving changes in products in *creative industries* (or changes in the aesthetic features of products in other industries).

Soft skills
A cluster of behavioral traits, personal habits and use of language that affect interaction with individuals and groups. Includes persuasive communications, creativity, strategic thinking and conflict resolution.

Springboards
A form of wishful thinking and expression used to assist in the process of idea generation. Includes "I wish" and "How to" statements and the application of images, connections, analogies and paraphrasing. (See: *Synectics®*)

Storyboard method
A form of idea development that goes beyond brainstorming methods; more organized than brainstorming. Enables a group to visualize, think and collaborate on a complex challenge. The collaboration involves creation of a structured, visual picture of how a solution might be implemented.

Suspended judgment
The ability of innovators to reserve their evaluation (and judgment) of new and emerging ideas. The goal in suspended judgment is to give idea generators an opportunity to develop their ideas into more concrete concepts before subjecting them to critical analysis and decisions.

Synectics®
An approach to innovation and problem-solving that stimulates new modes of thought of which the participant is generally unaware. Attempts to discover new and surprising solutions in ideation sessions by combining a structured approach to innovation with the spontaneous technique of brainstorming.

More complicated than brainstorming due to many steps, idea triggers and unique terminology; originally developed by the consulting firm Synectics, Inc., George Prince, William Gordon, and others. (See: *In/out listening, task headlines, springboards, threshold of acceptability*)

Systems thinking
An approach to planning and problem solving that treats complex challenges as parts of an overall system, rather than focusing on individual outcomes; based on the concept that component parts of a system are best understood in the context of relationships with each other and with other systems.

Tacit knowledge
Information and knowledge that an individual develops over time, often as a specialized skill, ability or method; may be difficult to articulate. Usually not documented or recorded. Can sometimes be demonstrated or taught to others or transferred through observation. (See: *Codified knowledge*)

Task headline
A carefully crafted one-sentence headline statement of what you expect a group to work on during an innovation session.

> "Unless there is articulated agreement by the group concerning what approach they are going take to the problem, there is confusion, because everyone works in his or her own, exclusive framework."
> Prince, p.26

Technology transfer
A process and system for sharing knowledge, specialized skills, technologies, software, manufacturing methods, and processes between and among corporate entities, universities, government organizations.

Sometimes used to commercially exploit research and new product concepts through licensing, joint ventures, development agreements and other partnerships designed to reduce risks and costs and share in the value generated.

> "Transfer = Transmission + Absorption
> ... if knowledge is not absorbed, it has not been transferred."
> Davenport and Prusak, p.101

Thought leadership
The intersection of knowledge management and marketing communications to demonstrate leadership in topics of interest to customers, partners, suppliers, other stakeholders and opinion leaders.

Threshold of acceptance
The point on an idea spectrum between acceptable and unacceptable where an idea or concept becomes usable in the minds of decision makers. Through the *itemized response* technique, a rough (and perhaps unattractive) idea is modified to overcome its uncertainties and risks and vault over the threshold. (See: *Synectics®*)

Transformation value model
A process model to assist planners with complex problems involving how to accomplish major changes in government and business organizations. Provides a dialogue for planners and promotes the combination of techniques from innovation management with strategic tools and objectives.

Triple helix
A model of university-industry-government interactions, based on theories that cooperation of these previously independent spheres is key to innovation and economic development in knowledge-based societies. The model suggests that when hybrid entities such as technology transfer departments in universities and government R&D labs interact with venture capitalists, entrepreneurs and technology-driven firms, the result is higher innovation capacity, value creation and economic growth.

Trust equation
A conceptual model to assist innovation teams in communicating and collaborating on the issues of risk-taking, knowledge sharing and role clarity.

Breakthrough! Innovation Management in Practice

Type I versus Type II errors
Types of errors encountered in a decision-making process. A Type I error, also known as a "false positive", occurs when we observe or perceive a difference when in actual fact there is none. A Type II error, also known as a "false negative" is an error of failing to observe a difference when in truth there is one. (Based on the work of statisticians Neyman and Pearson, 1928) For the innovation management professional, a Type II error is an error of excessive skepticism regarding an attractive opportunity.

Value creation
An activity or process that creates new assets, growth, knowledge or other objects of strategic value through learning, acquisition and technology transfer. Usually includes activities for systematizing the company's process for knowledge creation. (See: *Knowledge investment*)

Vertical thinking
An approach to decision-making and problem-solving that depends on an analytical, judgmental and usually sequential process (See: *Lateral thinking*)

Visionary
Someone who provides conceptual views and ideas on future opportunities. Usually skilled in describing future scenarios and sketching how a favorable outcome can become reality. Believes strongly in the innovation team's ability to control its process and destiny.

Weak signals
Subtle changes and discontinuities in an industry or business environment; usually discovered in market, competitive or technology trends. Detecting weak signals often serves as the *entry point* for innovation projects and idea generation activities. (See: *Entry point*)

White space mapping
A visual tool to help groups search for, discover and articulate new opportunities, typically in underserved or unserved markets—and perhaps in markets outside the core business. The white space is an area where products and services do not currently exist.

Working control
Having significant control of the direction, resources and content of an innovation project or meeting.

> "There are appropriate roles for each participant in a meeting. If these are understood and adhered to, the probability of success is substantially increased." Prince, p. 53

Appendix C–References

Anderson, Stuart (2016), "Immigrants and Billion-Dollar Startups." NFAP Policy Brief, National Foundation for American Policy

Arundel, Anthony; Bordoy, Catalina; and Kanerva, Minna, "Neglected Innovators: How Do Firms That Do Not Perform R&D Innovate?" Results of an analysis of the Innobarometer 2007 survey No. 215. Pro Inno Europe INNO-Metrics Thematic Paper, March 2008

Anthony, Scott A. (2009), *The Silver Lining: An Innovation Playbook for Uncertain Times.* Harvard Business School Press

Baldwin, Carliss Y. and Clark, Kim B (2006), "Architectural Innovation and Dynamic Competition: The Smaller Footprint Strategy." Working paper, Version 1.0, August 2006

Bieler, Dan (2015), "The Future of Telcos Remains Precarious." Forrester [blog post]. Available online:
http://blogs.forrester.com/dan_bieler/15-08-19
the_future_of_telcos_remains_precarious (accessed on 29 June 2017)

Bettinghaus, Erwin P. and Cody, Michael J. (1987), *Persuasive Communications.* Harcourt Brace Jovanovich College Publishers

Bristow, G.; Pill, M.; Davies, R.; and Drinkwater, S. (2011), "Stay, leave or return? Patterns of Welsh graduate mobility." *People, Place and Policy Online*, Vol. 5, No. 3, pp. 135-148 (as cited in Gkikas et. al., 2013)

Brueck, Hilary (2016), "Immigrants Are Driving Billion-Dollar Startups, Study Finds." FORTUNE. Available online at http://fortune.com/2016/03/18/billion-dollar-startup-founders-immigrants-study-nfap/ (accessed on 11 July 2017)

Burt, Ronald S. (2003), "Structural Holes and Good ideas." University of Chicago, pre-print of article for the *American Journal of Sociology*

Burt, Ronald S. (2003), "Social Origins of Good Ideas." Draft manuscript, University of Chicago

Chesbrough, Henry William (2003), *Open Innovation: The New Imperative for Creating and Profiting from Technology.* Harvard Business School Press

Breakthrough! Innovation Management in Practice

Chesbrough, Henry (2006), *Open Business Models: How to Thrive in the New Innovation Landscape*. Harvard Business School Press

Chourabi, Hafedh; Nam, Taewoo; Walker, Shawn; Gil-Garcia, J. Ramon; Mellouli, Sehl; Nahon, Karine; Pardo, Theresa A.; and Scholl, Hans Jochen (2012). "Understanding smart cities: An integrative framework." *IEEE Computer Society*, 978-0-7695-4525-7/12, DOI 10.1109/HICSS.2012.615, presented at 45th Hawaii International Conference on System Sciences

Christensen, Clayton (1997), *The Innovator's Dilemma: When New Technologies Cause Great Firms to Fail*. Harvard Business School Press

Coughlin, Michael S. and Younger, Mark (2004), "Measuring Value Creation." Accenture | The Postal Project (Volume One)

Dalziel, Margaret (2005), "The Impact of Industry Associations." School of Management, University of Ottawa, in *Innovation: Management, Policy and Practice*, eContent Management Pty Ltd, Queensland, Australia.

Davenport, Thomas H. and Prusak, Laurence (1998), *Working Knowledge: How Organizations Manage What They Know*. Harvard Business School Press

Day, George S., Schoemaker, Paul J. H. with Gunther Robert E. (2000), *Managing Emerging Technologies*. The Wharton School, John Wiley & Sons, Inc.

De Bono, Edward (1990), *Lateral Thinking: Creativity Step by Step*. Harper & Row, Publishers, Perennial Library Edition

Dickgreber, Florian; Campanini, Claudio; Grabowski, Soeren; Sorenson, Thomas (2015), "The Future of Telecom Operators in Europe," AT Kearney. Available online: https://www.atkearney.com/communications-media-technology/ideas-insights/the-future-of-telecom-operators-in-europe (accessed on 1 July 2017)

Diedrichs, Eva; Engel, Kai; and Wagner, Kristina, "European Innovation Management Landscape." Europe INNOVA paper No. 2. First edition, November 2006. © IMP³rove European Coordination Team

Dodgson, Mark and Gann, David (2010), *Innovation: A Very Short Introduction*. Oxford University Press

Drucker, Peter F. (1985), *Innovation and Entrepreneurship*. Harper & Row

Fagerberg, Jan; Mowery, David C. and Nelson, Richard D (2005), *The Oxford Handbook of Innovation*. Oxford University Press

Fisher, Anne (2014), "Why most innovations are great, big failures." Fortune. Available online: http://fortune.com/2014/10/07/innovation-failure/ (accessed on 16 July 2017)

Ghoshal, S. and Bartlett, C. (1988), "Creation, adoption, and diffusion of innovations by subsidiaries of multinational corporations." Journal of International Business Studies, 19.3, 365-389

Gilbert, Clark and Bower, Joseph L. (2002), "Disruption: The Art of Framing." Harvard Working Knowledge for Business Leaders

Glasco, Jon (2017), "Building Trust in Smart Cities: The Importance of Communications, Clarity and Civic Engagement" [unpublished manuscript, available at researchgate.com] © 2017 by Jon Glasco

Glasco, Jon (2013), *The Evolving Language of Innovation.* [self-published booklet] Printed by CreateSpace, An Amazon.com Company © 2013 by Jon Glasco

Glasco, Jon E. (2012), Web 2.0 and government transformation: How e-government and social media contribute to innovation in public services. [chapter] In Ed Downey and Matt Jones (Eds.), *Public Service, Governance and Web 2.0 Technologies: Future Trends in Social Media*, (pp. 201-222)

Glasco, Jon (2009), "Becoming a Top EU Research Nation: A National Thought Leadership Program for the Czech Republic." [unpublished white paper, available at researchgate.com] © 2009 by Jon Glasco

Glasco, Jon (2000), "Tips for Successful Written Agency Communications." PA Times, American Society of Public Administration, Vol. 23 No. 10

Gkikas, Aineias; Jones-Evans, Dylan; MacKenzie, Niall G. (2013), "Innovation performance of SMEs within an uncompetitive regional economy: the case of Wales." Paper presented at the 58th Annual International Council for Small Business World Conference, Ponce, Puerto Rico, 20-23 June 2013

Goldsmith, Stephen and Crawford, Susan (2014). *The Responsive City: Engaging Communities Through Data-Smart Governance*. Josey-Bass (A Wiley Brand), John Wiley & Sons

Breakthrough! Innovation Management in Practice

Grove, Andrew (1998), *Only the Paranoid Survive*. Profile Books Ltd., London; First published in Great Britain by HarperCollinsBusiness (1997). Copyright © Andrew Grove 1996

Hage, Jerald and Hollingsworth, J. Rogers (2000), "A Strategy for the Analysis of Idea Innovation Networks and Institutions." Organization Studies, EBSCO Publishing (www.bsos.umd.edu/socy/centerforinnovation)

Holt, Alex (2016), "The Future of Telco: Adapt to Thrive." © 2016 KPMG LLP

Hunt, Matt (2015), "Innovation and Failure - Understanding the Impact of Risk to Employees and the Organization." InnoChat. Available online at http://innochat.com/innochats/date/2015-03-19/innovation-and-failure-understanding-impact-to-employee-and-organizations (accessed on 16 July 2017)

International Economic Development Council (IEDC), "Economic Development Reference Guide"

Jaruzelski, Barry; Dehoff, Kevin and Bordia, Rakesh (2006), "Smart Spenders: The Global Innovation 1000." Booz Allen Hamilton Inc.

Johnson, Steven (2011), "Where Good Ideas Come From: The Natural History of Innovation." Riverhead Books

Jones, Ian (2016), "Improving the Economic Performance of Wales: Existing Evidence and Evidence Needs." Public Policy Institute for Wales. Copyright © 2016 Queen's Printer and Controller of HMSO

Kaner, Sam with Lind. L., Toldi, C., Fisk, S., and Berger, D. (2007), *Facilitator's Guide to Participatory Decision-Making*. Jossey-Bass, A Wiley Imprint. San Francisco, CA. Copyright © 2007 by Community At Work

Keeley, Larry; Pikkel, Ryan; Quinn, Brian; and Walters, Helen (2013), *Ten Types of Innovation: The Discipline of Building Breakthroughs*. John Wiley & Sons, Inc. Copyright © Deloitte Development LLC

Jones-Evans, D. and Bristow, G. (2010), "Innovation and the Objective 1 Programme I Wales: Lessons for the Convergence Fund.2 Paper presented to the Enterprise and Learning Committee, National Assembly for Wales, March 18th (as cited in Gkikas et. al., 2013)

Kelley, Tom (2007), General Manager IDEO, "The Ten Faces of Innovation." Lecture (based on his book) presented at "Dia L' Emprenedor," Barcelona, Spain, March 2007

Koestler, Arthur (1990), *The Act of Creation*. Viking Penguin, New York, NY

Komninos, Nicos; Tsarchopoulos, Panagiotis; and Kakderi, Christina (2014). New services design for smart cities: A planning roadmap for user-driven innovation. *URENIO Research*, Aristotle University of Thessaloniki

Koppett, Kat (2002), *Training Using Drama: Successful Development Techniques from Theatre and Improvisation*. Kogan Page Publishers

Lee, Neil (2005), "Ideopolis: Knowledge Cities | A Review of Quality of Life Measures." UK Work Foundation

Levitt, Theodore (1986), *Marketing Imagination*. The Free Press, a Division of Macmillan, Inc.

Leydesdorff, Loet and Guoping, Zeng (1996), "University-Industry-Government Relations in China: An Emergent National System of Innovations." Science & Technology Dynamics, University of Amsterdam and Center of Science, Technology and Society, Tsinghua University, Beijing

Luecke, Richard and Katz, Ralph (2003), *Managing Creativity and Innovation*. Boston, MA: Harvard Business School Press

Lundvall, Bengt-Åke (1992), *National Innovation Systems: Towards a Theory of Innovation and Interactive Learning*. London: Pinter Publishers

Mader, Isabella (2016). The new social contract: From representative to participative democracy (4.0). Excellence Institute. [Blog post] Available online: http://www.excellence-institute.at/en/the-new-social-contract-from-representative-to-participative-democracy-4-0/ (accessed on 17 June 2017)

Maastricht Economic Research Institute on Innovation and Technology (MERIT) and the Joint Research Center of the EC (2006), "European Innovation Scorecard 2006: Comparative Analysis of Innovation Performance." [http://www.proinno-europe.eu/doc/EIS2006_final.pdf]

Meffert, Jürgen and Mohr, Nick (2017), "Overwhelming OTT: Telcos' Growth Strategy in a Digital World," McKinsey & Company. Available online: http://www.mckinsey.com/industries/telecommunications/our-insights/overwhelming-ott-telcos-growth-strategy-in-a-digital-world

MRSC Local Government Services (2016), *Communication and Citizen Engagement Techniques*. Available online: http://mrsc.org/Home/Explore-Topics/Governance/Citizen-Participation-and-Engagement/Communication-and-Citizen-Participation-Techniques.aspx (accessed on 30 March 2017)

155

Mowery, David C. (1998), "The changing structure of the US national innovation system: Implications for international conflict and cooperation in R&D policy." Elsevier Research Policy. Copyright © 1998 Elsevier Science B.V.

Neckermann, Lukas with Smedley, Tim (2017), *Smart Cities, Smart Mobility: Transforming the Way We Live and Work."* Matador [Kindle edition]

Nolan, Vincent (1987), *The Innovator's Handbook: The Skills of Innovative Management.* Sphere Books Limited, The Penguin Group.

OECD (2017), *Embracing Innovation in Government: Global Trends.* World Government Summit | Dubai, United Arab Emirates, 12-14 February 2017

OECD (2014), *Risk Management and Corporate Governance.* OECD Publishing

OECD and Eurostat (2005), *Oslo Manual: Proposed Guidelines for Collecting and Interpreting Technological Innovation Data.* OECD Publishing

Osterwalder, Alexander and Pigneur, Yves (2010), *Business Model Generation.* John Wiley & Sons, Inc.

Payne, Mark (2014), *How to Kill a Unicorn: How the World's Hottest Innovation Factory Builds Bold Ideas That Make It to Market.* Crown Business (as cited in Fisher, 2014)

Pollock, Timothy G., Porac, Joseph F., and Wade, James B. (2004), "Constructing Deal Networks: Brokers as Network Architects in the US IPO Market and Other Examples." *Academy of Management Review,* Vol. 29, No. 1, 50-72.

Powell, Walter W. and Grodal, Stine (2005), "Networks of Innovators" in J. Fagerberg, D. C. Mowery and R. R. Nelson (Eds.) *The Oxford Handbook of Innovation,* Oxford University Press, pp. 56–85.

Prince, George (1970), *The Practice of Creativity.* Collier Books, A Division of Macmillan Publishing Co., Inc.

Rongping, Dr. Phil. Mu (2004), "Capability Building for Indigenous Innovation and Economic Development." Institute of Policy & Management, Chinese Academy of Sciences

Royce, William (1979), *Generating Strategic Alternatives.* SRI International: Business Intelligence Program

Saaty, Thomas (1980), *The Analytic Hierarchy Process*. McGraw Hill

Saxenian, A. (1994), *Regional Advantage: Culture and Competition in Silicon Valley and Route 128*. Harvard University Press, Cambridge, MA

Sernack, Janet (2016), "Eight reasons why innovation is important to businesses today." ImagineNation [innovation blog] Available online at http://www.imaginenation.com.au/innovation-blog/8-reasons-innovation-important-businesses-today/ (accessed on 8 July 2017)

Schoemaker, Paul J.H. and Mavaddat, V. Michael (2000), "Scenario Planning for Disruptive Technologies" in G. S. Day, P. J. H. Schoemaker and R. E. Gunther (Eds.) *Managing Emerging Technologies*. The Wharton School, John Wiley & Sons, Inc., pp. 206-241

Shukla, Amitabh (2017), "What is innovation? Why innovation is important?" [guest post] Available online: http://www.paggu.com/getting-into-roots/what-is-innovation-why-innovation-is-important/ (accessed on 6 July 2017)

Simon, Herbert A. (1985), "What We Know About the Creative Process" in R. L. Kuhn (Ed.), *Frontiers in Creative and Innovative Management*. Cambridge, MA. Ballinger Publishing Company, pp. 3-22

Smilor, Raymond W. (1996), "Entrepreneurship and Philanthropy." a presentation to the Kellogg-Kaufman Aspen Seminar on Philanthropy, Aspen Institute

Smutniak, John (2004), "Living Dangerously: A Survey of Risk." The Economist Newspaper Limited (print edition), January 22, 2004

Staack, Volker and Cole, Branton (2017), "Reinventing innovation: Five findings to guide strategy through execution." PwC | Key insights from PwC's Innovation Benchmark

The Economist (2004), "Be Prepared." The Economist Newspaper Limited (print edition), January 22, 2004

Toppeta, Donato (2010), "The Smart City Vision: How Innovation and ICT Can Build Smart, Liveable, Sustainable Cities." The Innovation Knowledge Foundation [THINK! Report 005/2010]

Torres, Joan (2017), "Smart Cities: Past, Present and Future." LabCities. Available online: http://www.labcities.com/smart-cities-past-present-future/?utm_content=buffer6aba8&utm_medium=social&utm_source=twitter.com&utm_campaign=buffer (accessed on 7 July 2017)

UK Department of Trade and Industry (2006), "Succeeding Through Innovation: Turning Ideas Into Profit." A joint publication of British Chambers of Commerce (BCC), Institute of Directors (IOD), Federation of Small Businesses (FSB), and Confederation of British Industry (CBI)

US Department of Commerce (2012), "The Competitiveness and Innovative Capacity of the United States." In consultation with the National Economic Council

Valéry, Nicholas (1999), "Industry Gets Religion." The Economist Newspaper Limited, London, February 18, 1999

Wagner, Kristina (2007), "Characteristics of Leading Innovators." A. T. Kearney. Presentation at INNO-Views Policy Workshop, Eindhoven

Watson, Richard (2006), "Expand Your Innovation Horizons." Fast Company (www.fastcompany.com); Mansueto Ventures LLC

Welsh Assembly Government (2003), "Management and Innovation." Wales Management Council, (www.crc-wmc.org.uk), William Battle Associates Ltd

Welsh Assembly Government (circa 2003), "Wales for Innovation: An Innovation Action Plan." The Innovation Branch (IAP Consultation) Economic Development Department

Wenger, E. (1998), *Communities of Practice*. Cambridge University Press, New York

Wilford, John Noble (1992), *The Mysterious History of Columbus*. Vintage Books, A Division of Random House, Inc., New York

Worldbank (2006), "Korea as a Knowledge Economy: Evolutionary Process and Lessons Learned." The World Bank, Washington, DC

Worldbank (2000), "Republic of Korea: Transition to a Knowledge-Based Economy." The World Bank, East Asia and Pacific Region [Report No. 20346-KO, June 29, 2000]

Author Profile

Jon Glasco is a freelance consultant and author with knowledge of innovation management principles derived from 25+ years in Fortune 500 companies, start-up enterprises, government agencies and multinational ventures.

Since the 1990s, Jon has held key roles as a planning consultant and technical writer for clients in Austria, Canada, the Czech Republic, Germany, Mexico, Spain, Switzerland, the UK and US.

His range of industry experience and knowledge includes:

- Telecommunication services, mobile and broadband networks
- Intelligent mobility, public transport and rail vehicle manufacturing
- Professional services and management consulting
- Fire safety, R&D and design engineering
- High-tech sectors (fiber optics, electronics, multimedia)

His clients and employers have included AirTouch International (acquired by Vodafone), A. T. Kearney, Alstom Transport, Chevron, Electronic Data Systems, Ernst & Young, GE Security, Jazztel, ONE GmbH, Pacific Bell, Sprint Communications, Transport Systems Catapult, United Technologies, US Department of Energy, and the Welsh Development Agency.

Jon is the author of more than 20 white papers, articles, and thought leadership publications on innovation, strategy and business communications. He holds an MBA degree and a Bachelor of Science in Electrical Engineering.

Email: jon.glasco@gcaconsulting.com
Twitter: @JonGlasco
LinkedIN: www.linkedin.com/in/jeglasco3

"It isn't all over; everything has not been invented; the human adventure is just beginning."

Gene Roddenberry